Justin Lee

# Morphogenetic Evolvable Hardware

Justin Lee

# Morphogenetic Evolvable Hardware

Biological Models for Generating Electronic Circuits on FPGAs

VDM Verlag Dr. Müller

# Imprint

Bibliographic information by the German National Library: The German National Library lists this publication at the German National Bibliography; detailed bibliographic information is available on the Internet at http://dnb.d-nb.de.

Any brand names and product names mentioned in this book are subject to trademark, brand or patent protection and are trademarks or registered trademarks of their respective holders. The use of brand names, product names, common names, trade names, product descriptions etc. even without a particular marking in this works is in no way to be construed to mean that such names may be regarded as unrestricted in respect of trademark and brand protection legislation and could thus be used by anyone.

Cover image: www.purestockx.com

Publisher:
VDM Verlag Dr. Müller Aktiengesellschaft & Co. KG , Dudweiler Landstr. 125 a, 66123 Saarbrücken, Germany,
Phone +49 681 9100-698, Fax +49 681 9100-988,
Email: info@vdm-verlag.de

Zugl.: Brisbane, Queensland University of Technology, Diss., 2006

Produced in USA and UK by:
Lightning Source Inc., La Vergne, Tennessee, USA
Lightning Source UK Ltd., Milton Keynes, UK
BookSurge LLC, 5341 Dorchester Road, Suite 16, North Charleston, SC 29418, USA

ISBN: 978-3-639-05716-4

# Contents

**Acronyms** xi

**Acknowledgements** xiii

**1 Introduction** 1
  1.1 The Importance of Evolvable Hardware . . . . . . . . . . . . . . 1
  1.2 What is Evolvable Hardware . . . . . . . . . . . . . . . . . . . 3
  1.3 The Main Problems to be solved in EHW . . . . . . . . . . . . 4
  1.4 Motivation For This Research . . . . . . . . . . . . . . . . . . 5
  1.5 Aim of the Project . . . . . . . . . . . . . . . . . . . . . . . . 6
  1.6 Outline of Book . . . . . . . . . . . . . . . . . . . . . . . . . 7

**2 Evolvable Hardware Background** 11
  2.1 Definition . . . . . . . . . . . . . . . . . . . . . . . . . . . . 11
    2.1.1 Evolutionary Computation Overview . . . . . . . . . . 13
    2.1.2 Intrinsic vs. Extrinsic . . . . . . . . . . . . . . . . . . 17
  2.2 Applications . . . . . . . . . . . . . . . . . . . . . . . . . . . 18
    2.2.1 Online Adaptation . . . . . . . . . . . . . . . . . . . . 19
    2.2.2 EHW Robot Controllers . . . . . . . . . . . . . . . . . 21
    2.2.3 Advantages of EHW Controllers . . . . . . . . . . . . . 24
  2.3 Abstraction Level and Platform . . . . . . . . . . . . . . . . . 26
    2.3.1 Abstraction Level . . . . . . . . . . . . . . . . . . . . 26
    2.3.2 Field Programmable Gate Arrays . . . . . . . . . . . . 27
    2.3.3 Analogue Platforms . . . . . . . . . . . . . . . . . . . 30
    2.3.4 Other Architectures . . . . . . . . . . . . . . . . . . . 32
  2.4 The Xilinx Virtex FPGA . . . . . . . . . . . . . . . . . . . . 33
    2.4.1 Virtex Architecture . . . . . . . . . . . . . . . . . . . 34
    2.4.2 Virtex Configuration . . . . . . . . . . . . . . . . . . . 36
    2.4.3 Configuration with JBits . . . . . . . . . . . . . . . . . 36
  2.5 Problems in EHW . . . . . . . . . . . . . . . . . . . . . . . . 37
    2.5.1 Circuit Validity Issues . . . . . . . . . . . . . . . . . . 37

# Contents

2.5.2    Scalability . . . . . . . . . . . . . . . . . . . . . . . . 41

2.6   Summary . . . . . . . . . . . . . . . . . . . . . . . . . . 44

**3   Morphogenesis Background**     **45**

3.1   Properties of Morphogenesis . . . . . . . . . . . . . . . . 45

    3.1.1    Scalability of Morphogenesis . . . . . . . . . . . . . 46

    3.1.2    Genetic Robustness and Phenotype Legality . . . . . . 47

    3.1.3    Aiding Evolutionary Search with Neutral Pathways . . . 48

    3.1.4    Modularity of Development for Incremental Evolution . . 50

3.2   Biological Background of Morphogenesis . . . . . . . . . . . 52

    3.2.1    The Cell . . . . . . . . . . . . . . . . . . . . . . . . 52

    3.2.2    Developmental Processes . . . . . . . . . . . . . . . 53

       3.2.2.1    Growth and Cell Division . . . . . . . . . . 54

       3.2.2.2    Pattern Formation . . . . . . . . . . . . . 54

       3.2.2.3    Differentiation . . . . . . . . . . . . . . . 55

       3.2.2.4    Morphogenesis . . . . . . . . . . . . . . . 55

    3.2.3    Gene Expression . . . . . . . . . . . . . . . . . . . 56

       3.2.3.1    Operon Model of Gene Expression . . . . . . 59

    3.2.4    Inter-Cellular Signaling . . . . . . . . . . . . . . . . 61

3.3   Morphogenesis in EHW and EC . . . . . . . . . . . . . . . 63

    3.3.1    L-Systems . . . . . . . . . . . . . . . . . . . . . . . 65

    3.3.2    L-Systems in EHW . . . . . . . . . . . . . . . . . . 69

    3.3.3    Cell-based Models for EHW . . . . . . . . . . . . . . 71

    3.3.4    Computational Models of Gene Expression . . . . . . . 76

3.4   Summary . . . . . . . . . . . . . . . . . . . . . . . . . . 78

**4   Designing a Morphogenetic EHW System**     **81**

4.1   Choosing Resources for EHW . . . . . . . . . . . . . . . . 82

4.2   Mapping Biology to Hardware Structures . . . . . . . . . . . 84

4.3   Developmental Processes . . . . . . . . . . . . . . . . . . . 87

    4.3.1    Growth and Cell Division . . . . . . . . . . . . . . . 87

    4.3.2    Pattern Formation and Differentiation . . . . . . . . . 88

    4.3.3    Morphogenesis . . . . . . . . . . . . . . . . . . . . 88

4.4   Gene Expression Model . . . . . . . . . . . . . . . . . . . 89

    4.4.1    Chromosome Encoding . . . . . . . . . . . . . . . . 91

    4.4.2    Codon-based Genetic and Protein Bind Signature Codes   93

    4.4.3    Example of Gene Expression Driven by FPGA Configuration . . . . . . . . . . . . . . . . . . . . . . . . . 95

4.5   Genetic Operators . . . . . . . . . . . . . . . . . . . . . . 97

    4.5.1    Homologous Crossover . . . . . . . . . . . . . . . . 98

    4.5.2    Base Mutation . . . . . . . . . . . . . . . . . . . . 99

    4.5.3    Frame Shift and Inversion Mutations . . . . . . . . . . 100

4.6   Summary . . . . . . . . . . . . . . . . . . . . . . . . . . 101

**5  Morphogenetic EHW Implementation**                                    **105**
  5.1  Setup . . . . . . . . . . . . . . . . . . . . . . . . . . . . 107
  5.2  Evolution . . . . . . . . . . . . . . . . . . . . . . . . . . 108
  5.3  Chromosome Preprocessing . . . . . . . . . . . . . . . . . 110
  5.4  Morphogenesis . . . . . . . . . . . . . . . . . . . . . . . 115
  5.5  Hardware Morphogenesis Interface . . . . . . . . . . . . . 118
  5.6  Circuit Evaluation . . . . . . . . . . . . . . . . . . . . . 119
  5.7  Summary . . . . . . . . . . . . . . . . . . . . . . . . . . 119

**6  Experiments with Evolving Circuit Structure**                         **121**
  6.1  Resource Allocation . . . . . . . . . . . . . . . . . . . . . 122
  6.2  Direct Encoding Chromosome Structure . . . . . . . . . . 123
  6.3  Fitness Evaluation . . . . . . . . . . . . . . . . . . . . . 123
  6.4  Signal Routing Experiments . . . . . . . . . . . . . . . . 124
      6.4.1  Experiments in Scaling Circuit Size . . . . . . . . . . 125
      6.4.2  Experiments in Scaling Circuit Complexity . . . . . . 131
      6.4.3  Summary of Routing Experiment Results . . . . . . . 133
  6.5  Summary . . . . . . . . . . . . . . . . . . . . . . . . . . 134

**7  Simulated Secondary Developmental Mechanisms**                        **137**
  7.1  Useful Biological Mechanisms Requiring Simulated Molecules . . 138
  7.2  Details of The Implementation of Simulated Molecules . . . . . 140
  7.3  Experiments . . . . . . . . . . . . . . . . . . . . . . . . . 143
  7.4  Discussion . . . . . . . . . . . . . . . . . . . . . . . . . . 147
  7.5  Summary . . . . . . . . . . . . . . . . . . . . . . . . . . 149

**8  Experiments with Evolving Circuit Functionality**                     **151**
  8.1  LUT Encodings . . . . . . . . . . . . . . . . . . . . . . . 152
  8.2  Experimental Setup . . . . . . . . . . . . . . . . . . . . . 153
  8.3  Fitness Evaluation . . . . . . . . . . . . . . . . . . . . . 158
  8.4  Experimental Runs . . . . . . . . . . . . . . . . . . . . . 159
  8.5  Hand-Crafted Solution . . . . . . . . . . . . . . . . . . . 161
  8.6  Experiments Isolating the Causes of Failure . . . . . . . . 163
  8.7  Reported Work Evolving Adders on a Virtex . . . . . . . . 166
  8.8  Summary . . . . . . . . . . . . . . . . . . . . . . . . . . 168

**9  Estimation of Problem Solvability**                                   **171**
  9.1  Measuring EHW Circuit Complexity . . . . . . . . . . . . 171
      9.1.1  Measuring Circuit Complexity . . . . . . . . . . . . 172
      9.1.2  Measuring Fitness Feedback . . . . . . . . . . . . . 174
  9.2  Complexity Calculation Results . . . . . . . . . . . . . . . 176
  9.3  Solvability Heuristic . . . . . . . . . . . . . . . . . . . . . 177
  9.4  Summary . . . . . . . . . . . . . . . . . . . . . . . . . . 180

Contents

**10 Conclusion**                                                   **183**
10.1 Summary . . . . . . . . . . . . . . . . . . . . . . . . . . . 183
10.2 Project Outcomes . . . . . . . . . . . . . . . . . . . . . . . 189
10.3 Discussion and Further Work . . . . . . . . . . . . . . . . . 191
    10.3.1 Platform Issues in Adder Experiments . . . . . . . . . 191
    10.3.2 Issues in Morphogenetic Scalability . . . . . . . . . . . 193
    10.3.3 Future Work on the Solvability Heuristic . . . . . . . . 196
    10.3.4 Hardware Implementation . . . . . . . . . . . . . . . . 196

**A Exploratory Experiments**                               **199**
A.1 Morphogenesis vs Direct Encoding . . . . . . . . . . . . . . 200
A.2 Comparison of 5x5 and 8x8 Morphogenesis Runs . . . . . . . 205
A.3 Simulated Transcription Factor Experiments . . . . . . . . . . 212
    A.3.1 MG vs TF Initial Test . . . . . . . . . . . . . . . . 212
    A.3.2 MG vs TF Test Rerun . . . . . . . . . . . . . . . . 213
    A.3.3 MG vs TF Results and Discussion . . . . . . . . . . . . 215
A.4 Speed Tests . . . . . . . . . . . . . . . . . . . . . . . . . . . 216
A.5 Sensitivity Analysis . . . . . . . . . . . . . . . . . . . . . . . 218
    A.5.1 Sensitivity of Evolutionary Parameters . . . . . . . . . 218
    A.5.2 Sensitivity of Morphogenetic Parameters . . . . . . . . 219
    A.5.3 Sensitivity to Anti-Gene-Bloat . . . . . . . . . . . . . 220
A.6 Summary of Results . . . . . . . . . . . . . . . . . . . . . . . 221

**B Experiment Complexity Calculations**                     **223**
B.1 Routing Experiment Calculations . . . . . . . . . . . . . . . . 223
    B.1.1 Calculating State Space . . . . . . . . . . . . . . . . 223
    B.1.2 Calculating Fitness Feedback . . . . . . . . . . . . . 224
    B.1.3 Calculating Problem Difficulty . . . . . . . . . . . . . 225
    B.1.4 Summary of Routing Experiment Calculations . . . . . 230
B.2 Adder Experiment Calculations . . . . . . . . . . . . . . . . . 230
    B.2.1 Calculating State Space . . . . . . . . . . . . . . . . 230
    B.2.2 Calculating Fitness Feedback . . . . . . . . . . . . . 231
    B.2.3 Calculating Problem Difficulty . . . . . . . . . . . . . 233
B.3 Experiment Calculation Summary . . . . . . . . . . . . . . . . 252

**C Genetic Code Details**                                        **255**
C.1 Resource Encoding . . . . . . . . . . . . . . . . . . . . . . . . 255
    C.1.1 Resource-Attribute Encoding . . . . . . . . . . . . . 255
    C.1.2 Resource Settings Encoding . . . . . . . . . . . . . . 258
C.2 Notes on Conversion to JBits Specification . . . . . . . . . . . 272
    C.2.1 Slim Subset for Signal Routing . . . . . . . . . . . . 272
    C.2.2 Unregistered Subset for One Bit Adder . . . . . . . . . 276
C.3 Notes on Direct Encoding . . . . . . . . . . . . . . . . . . . . 279

C.3.1   Direct Encoding for Signal Routing Experiments . . . . . 279
C.3.2   Direct Encoding for Adder Experiments   . . . . . . . . . 281

# Contents

# List of Figures

2.1   Outline of Evolutionary Algorithm . . . . . . . . . . . . . . . 14
2.2   Xilinx XC6200 Cell . . . . . . . . . . . . . . . . . . . . . . . 29
2.3   Virtex Structure . . . . . . . . . . . . . . . . . . . . . . . . . 34
2.4   Virtex Slice Structure . . . . . . . . . . . . . . . . . . . . . . 35
2.5   Virtex Slice Muxes . . . . . . . . . . . . . . . . . . . . . . . 38
2.6   Virtex CLB Routing . . . . . . . . . . . . . . . . . . . . . . 39
2.7   Direct Encoding EHW . . . . . . . . . . . . . . . . . . . . . 42

3.1   Structure of a Eukaryote Cell . . . . . . . . . . . . . . . . . 53
3.2   Structure of a Eukaryote Gene . . . . . . . . . . . . . . . . . 59
3.3   Structure of the *lac* Operon . . . . . . . . . . . . . . . . . 61
3.4   Morphogenetic EHW Encoding . . . . . . . . . . . . . . . . 72

4.1   MGEHW System Supported Slice Resources . . . . . . . . . . 83
4.2   MGEHW System Supported Routing . . . . . . . . . . . . . . 84
4.3   Example Cell/Protein to Hardware Correspondence . . . . . . . 87
4.4   Chromosome Regions . . . . . . . . . . . . . . . . . . . . . . 93
4.5   Example of Gene Expression Driven By Hardware State . . . . . 97
4.6   Initiation of Transcription with Polymerase Binding . . . . . . 98
4.7   Gene Transcription . . . . . . . . . . . . . . . . . . . . . . . 98
4.8   Gene Transcription (cont.) . . . . . . . . . . . . . . . . . . . 99
4.9   Gene Transcription Terminates . . . . . . . . . . . . . . . . . 99
4.10  Gene Transcription Repressed . . . . . . . . . . . . . . . . . 100
4.11  Homologous Crossover . . . . . . . . . . . . . . . . . . . . . 101
4.12  Single Base Mutations . . . . . . . . . . . . . . . . . . . . . 101
4.13  Frameshift Mutations . . . . . . . . . . . . . . . . . . . . . . 102
4.14  Inversion . . . . . . . . . . . . . . . . . . . . . . . . . . . . . 102

5.1   MGEHW Process Flow . . . . . . . . . . . . . . . . . . . . . 106
5.2   Chromosome to Morphogenetic Specification Process . . . . . 111
5.3   Cell Update Order . . . . . . . . . . . . . . . . . . . . . . . 116

6.1   Structure of S0G-Y Slim-subset Cell . . . . . . . . . . . . . . . 122
6.2   Single Signal Routing on a 5x5 CLB Matrix  . . . . . . . . . . 126
6.3   Mean and S.D. Generations Required to find Solution on Single
      Signal Routing  . . . . . . . . . . . . . . . . . . . . . . . . . . . 127
6.4   Mean and S.D. Chromosome Length for Solution on Single Sig-
      nal Routing . . . . . . . . . . . . . . . . . . . . . . . . . . . . . 128
6.5   Mean and S.D. Genes in Solution on Single Signal Routing . . . 129
6.6   Mean and S.D. Growth Steps Required for Solution on Single
      Signal Routing  . . . . . . . . . . . . . . . . . . . . . . . . . . . 130
6.7   Four Signal Routing on a 8x8 CLB Matrix . . . . . . . . . . . 132
6.8   Mean Maximum Fitness for Signal Routing on 8x8 CLBs . . . . 133

7.1   Cytoplasm Placement for Input layers . . . . . . . . . . . . . . 145
7.2   Cytoplasm Placement for Output layers . . . . . . . . . . . . . 145
7.3   Mean Maximum Fitness for TF vs MG Approaches on Signal
      Routing  . . . . . . . . . . . . . . . . . . . . . . . . . . . . . . . 146
7.4   Mean Maximum Fitness for TF vs MG Approaches on Multi
      Signal Routing  . . . . . . . . . . . . . . . . . . . . . . . . . . . 147
7.5   Mean Maximum Fitness for TF vs MG Approaches on Signal
      Routing  . . . . . . . . . . . . . . . . . . . . . . . . . . . . . . . 148

8.1   Layout for Adder Experiments . . . . . . . . . . . . . . . . . . 154
8.2   Structure of S0G-Y Logic Element Cell . . . . . . . . . . . . . 156
8.3   Maximum Fitness for Evolving 1-bit Adders  . . . . . . . . . . 161
8.4   Maximum Fitness for Evolving Adders with Modified Hamming  164

A.1   Mean and SD maximum fitness for 5x5 CLB matrix: MG vs GA 201
A.2   Mean and SD maximum fitness for 8x8 CLB matrix: MG vs GA 203
A.3   Mean max fitness for routing IO across 5x5 and 8x8 CLB matrix:
      MG vs GA . . . . . . . . . . . . . . . . . . . . . . . . . . . . . . 205
A.4   Mean maximum fitness for 5x5 CLB matrix MG and TF runs  . 216
A.5   Mean maximum fitness for 8x8 CLB matrix MG and TF runs  . 217

B.1   Shortest Paths on a 5x5 CLB Matrix  . . . . . . . . . . . . . . 226
B.2   Shortest Path on a Rx(C+4) CLB Matrix  . . . . . . . . . . . 228
B.3   Shortest Paths on a 8x8 CLB Matrix  . . . . . . . . . . . . . . 229

# List of Tables

3.1 The Genetic Code . . . . . . . . . . . . . . . . . . . . . . . . . . 57

6.1 Allocation of Resources to Logic Elements . . . . . . . . . . . . 122
6.2 Signal Routing Experiment Parameters . . . . . . . . . . . . . . 125
6.3 Summary of Signal Routing Experiment Sets . . . . . . . . . . . 134

7.1 Comparison of Morphogenetic Runs with and without Simulated TFs . . . . . . . . . . . . . . . . . . . . . . . . . . . . . . 146

8.1 Adder IO Connection Points . . . . . . . . . . . . . . . . . . . . 155
8.2 Available LUT Inputs . . . . . . . . . . . . . . . . . . . . . . . 157
8.3 Available CLB Outputs . . . . . . . . . . . . . . . . . . . . . . 157
8.4 Adder Experiment Parameters . . . . . . . . . . . . . . . . . . . 160
8.5 Preconfigured Adder Experiment Run Results . . . . . . . . . . 165
8.6 Fixed Adder Experiment Run Results . . . . . . . . . . . . . . . 167

9.1 Experiment Calculation Summaries in Base 2 Logs . . . . . . . 177
9.2 Adder Experiment calculated E and required E* ($\log_2$), for k=2 179

A.1 Signal Routing Experiment Results . . . . . . . . . . . . . . . . 203
A.2 Details of Fittest Evolved Chromosomes for Morphogenesis on
5x5 and 8x8 CLB Matrix . . . . . . . . . . . . . . . . . . . . . 204
A.3 Comparison of Morphogenetic Runs . . . . . . . . . . . . . . . 206
A.4 Resource Codon Allocations in Genetic Code . . . . . . . . . . 208
A.5 Results of Morphogenesis with TFs for Signal Routing . . . . . 213
A.6 Results of Morphogenesis with Longer Life TFs for Signal Routing 215
A.7 Comparison of Morphogenetic Runs with and without Simu-
lated TFs . . . . . . . . . . . . . . . . . . . . . . . . . . . . . . 215
A.8 Speed Tests For Routing Experiments Over 100 Generations . . 217
A.9 Sensitivity Analysis on Evolutionary Parameters . . . . . . . . 219
A.10 Sensitivity Analysis on Evolutionary Parameters (cont.) . . . . 219
A.11 Sensitivity Analysis on Evolutionary Parameters . . . . . . . . 220

A.12 Sensitivity Analysis for Anti Gene Bloat . . . . . . . . . . . . 220

B.1 Routing Experiment State Spaces . . . . . . . . . . . . . . . 224
B.2 Routing Experiment Fitness Feedback . . . . . . . . . . . . . 225
B.3 Routing Experiment Probabilities in Base 2 Logs . . . . . . . 230
B.4 Adder Experiment State Spaces in Base 2 Logs . . . . . . . . 231
B.5 Adder Experiment Fitness Feedback . . . . . . . . . . . . . . 234
B.6 Used LUT Boolean Truth Table Entries . . . . . . . . . . . . 235
B.7 Experiment Calculation Summaries in Base 2 Logs . . . . . . 253

C.1 Top Level Virtex Genetic Code . . . . . . . . . . . . . . . . 256
C.2 Top Level Virtex Bind Code . . . . . . . . . . . . . . . . . . 257
C.3 Resource-Attribute to JBits Correspondence . . . . . . . . . 258
C.4 LUT Active and Incremental Functions Operator Encoding . . . 260
C.5 Bus – Single Encoding . . . . . . . . . . . . . . . . . . . . . 262
C.6 OutToSingleBus Single Line Ordering . . . . . . . . . . . . . 262
C.7 Out To Single Dir Line Encoding . . . . . . . . . . . . . . . 263
C.8 Out To Single Dir Single Line Ordering . . . . . . . . . . . . 264
C.9 Slice Input Mux Encoding . . . . . . . . . . . . . . . . . . . 265
C.10 Slice Input Mux Line Encoding . . . . . . . . . . . . . . . . 265
C.11 Slice Mux Input Line Ordering . . . . . . . . . . . . . . . . . 267
C.12 Slice Output Mux Line Encoding . . . . . . . . . . . . . . . . 268
C.13 Slice Output Mux Line Ordering . . . . . . . . . . . . . . . . 268
C.14 SliceRAM Attribute-Functionality Encoding . . . . . . . . . . 269
C.15 SliceRAM Attribute/Setting to JBits . . . . . . . . . . . . . 270
C.16 TF Bind Sequence Encoding . . . . . . . . . . . . . . . . . . 272
C.17 Slim Subset SliceIn Line Number to Input Line Mapping . . . . 273
C.18 Slim Subset SliceToOut Out Bus Line and Line Number Mappings 274
C.19 Slim Subset OutToSingle Out Bus Line and Line Number Map-
    pings . . . . . . . . . . . . . . . . . . . . . . . . . . . . . . . 274
C.20 Slim Subset OutToSingle Out Bus Line and Line Number Map-
    pings . . . . . . . . . . . . . . . . . . . . . . . . . . . . . . . 275
C.21 Slim Subset LUTBitFN Mapping . . . . . . . . . . . . . . . . 276
C.22 Unregistered Subset SliceIn Input Line Ordering . . . . . . . . 277
C.23 Unregistered Subset OutToSingleBus Line Number Mappings . . 278
C.24 Unregistered Subset OutToSingleDir Line Number Mappings . . 279
C.25 Direct Encoding Slim Subset Out Bus Assignments and Settings 280
C.26 Direct Encoding Slim Subset OutToSingle PIPs . . . . . . . . 281
C.27 Direct Encoding Adder SliceIn Input Line Ordering . . . . . . 282
C.28 Direct Encoding Unregistered Subset OutToSingle PIPs . . . . . 283

# Acronyms

| | |
|---|---|
| BRAM | Block RAM |
| CLB | Configurable Logic Block |
| E# | Routing line SINGLE_EAST# (e.g. E2 = SINGLE_EAST2) |
| EHW | Evolvable Hardware |
| FPGA | Field Programmable Gate Array |
| GA | Genetic Algorithm (also direct encoding approach) |
| IO | Input/Output |
| IOB | Input/Output Block |
| LE | Logic Element (LUT-register pair in slice) |
| LUT | Look-Up-Table (Boolean function generator in logic element) |
| MG | Morphogenetic (approach) |
| MGEHW | Morphogenetic Evolvable Hardware |
| Mux | Multiplexor |
| N# | Routing line SINGLE_NORTH# (e.g. N2 = SINGLE_NORTH2) |
| PIP | Programmable Interconnection Point (in FPGA routing) |
| RAM | Random Access Memory |
| S# | Routing line SINGLE_SOUTH# (e.g. S2 = SINGLE_SOUTH2) |
| S0G-Y | CLB slice 0, logic element LUT G – slice output Y (and YQ) |
| S0F-X | CLB slice 0, logic element LUT F – slice output X (and XQ) |
| S1G-Y | CLB slice 1, logic element LUT G – slice output Y (and YQ) |
| S1F-X | CLB slice 1, logic element LUT F – slice output X (and XQ) |
| S.D. | Standard Deviation |
| TF | Transcription Factor (also simulated TF utilising approach) |
| W# | Routing line SINGLE_WEST# (e.g. W2 = SINGLE_WEST2) |

# Acknowledgements

*I dedicate this book to my maternal grandfather, retired engineer, G.P. Tuck, who introduced me to a wide range of engineering knowledge at a young age, started me off in electronics and computers, taught me the humility of knowledge, and provided both the nature (genes) and nurture (environment) that made this work possible.*

I would like to acknowledge my principal Ph.D. supervisor Joaquin Sitte, the leader of the Smart Devices Lab, for his long-term vision, trust, patience and support, which allowed me to follow the research tangent that has become the work presented in this book.

Frederic Maire and Hussein Abbass, also both require special mention for their unceasing support, advice and sharing of knowledge.

Man cannot live on research alone: food, clothes and shelter come in quite handy. For this, I'd like to thank Aster Wardhani and Ross Hayward for providing me with computer architecture tutoring, which kept me fed for over two years.

I would like to thank all my past and present colleagues in the Smart Devices Lab, the visiting Dipl. Ing. project students (particularly from the Circuit and Systems Group at Paderborn University), and others for sharing the days, nights, weekends and holidays in the research center. In particular, I would like to thank Dylan Muir, especially for letting me pick his brain on FPGA-related issues; Erik Berglund for Java assistance and Quake stress relief; Manix Au for stats advice and music that kept me sane; Chris Pohl for technical advice on the Raptor FPGA development board; Johan Berntsson for genetic algorithm tips; Rob Smith for Java and Latex formatting assistance; Maxim Mikhalsky for aiding with some experiment runs; Ross Brown for writing advice; Alex Campbell for Latex help and insights in complex systems; and lastly, Lawrence Tham and Michael Towsey who provided invaluable advice in molecular biology.

Ryan Easter, Liz Lipowitz and other present and past tech support and research admin staff also deserve thanks for their behind the scenes assistance.

Most importantly, I would like to thank all my friends and family, for all the big things and the little things that make life outside of my head worth living. There are too many to name individually in such a short space. So, to all the old Darwin mob, though most of us have left, thanks for staying in touch, even while I have had my head buried in this PhD. To all the Brissie gang, thanks for making me at home here, and all the coffee breaks, movies, dinners, and other escapes from study.

Some friends, I can single out for contributing directly to this work: Warrakun Mangrai, for the generous loan of several PCs to conduct experiments on, and along with his mother and sister Sree, helping me settle in Brisbane; John Simpson for supplying textbooks at bargain prices, and sharing accommodation with a domestic-duties negligent PhD student; Barry Arthur for assistance with Perl, and providing a place to sleep when time and money were in short supply; Jimmy Doukas for help with getting the final thesis bound; Piao Lee for advice on useful software tools, and for his skills in locating absolutely anything on the Internet; and Poh Ling Yap and Steve Lew for their time and generousity, and including an "extra wheel" on their holidays.

Last, but foremost, I'd like to thank my best friend Trieu Thai. I have Trieu to thank for many things, some directly related to this PhD, and most not. True friendship like good wine improves with age — my words run dry, but the wine still flows.

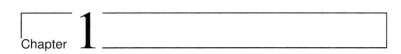

Chapter **1**

# Introduction

## 1.1 The Importance of Evolvable Hardware

During the Second NASA/DoD Workshop on Evolvable Hardware in July
2000, Dr. Zornetzer, the director of the NASA Ames Information Systems
and Technology Directorate, stated that "Maybe NASA's future will depend
on evolvable hardware." (Lohn, Stoica, Keymeulen, and Colombano 2001) He
then gave the following reasons why NASA is interested in evolvable hardware
(EHW or EH):

- Deep space and planetary exploration vehicles need robust system architectures that are dynamically reconfigurable.

- To explore unknown and unforeseeable environments, autonomous real-
  time adaptive robots are needed. An example of this is for probes sent
  into the saline oceans under Europa's ice cap.

- Triply redundant systems for robustness are impractical due to their
  extra weight, instead what is required is reconfigurability.

At the same workshop, Nikzad Toomarian of the Jet Propulsion Laboratory (JPL) stated that "EH is needed for deep space exploration in extreme
environments." (Lohn, Stoica, Keymeulen, and Colombano 2001) According to
Toomarian, various space missions, such as the Pluto express, with a flight time
of 8–9 years, and later interstellar explorations, need to emphasize long-term
survivability and evolvability. Furthermore JPL want future space hardware
systems based on nature's adaptability (EHW) that can adapt in seconds and
survive for over 100 years with low power and high intelligence.

**Key Strengths of Evolvable Hardware**

Evolvable hardware is able to achieve adaptation and fault tolerance in several manners. The most obvious of these is that when a given circuit ceases to

function as required, due to unexpected environmental effects for example, evolution can be continued until a new circuit is found that produces the desired behaviour under the current circumstances. In actual fact, for circuits that cannot be fully specified a priori, but where the desired behaviour is known, evolutionary approaches may be the only viable option. In conjunction with this, circuits can be evolved specifically to be adaptive and resilient to faults. Lastly, evolution is able to incorporate faulty components into the circuit, according to their actual behaviour, as opposed to their expected behaviour.

From this some of the key strengths of EHW in comparison with conventional electronics design can be seen. Specifically, EHW offers the possibility of generating circuits that are able to adapt to circumstances the designer cannot not foresee, offer robustness without requiring costly back up systems, and better utilise the available resources (in terms of space requirements or energy efficiency, for example).

This latter strength is due to the ability of EHW to produce circuits that are not constrained by human design methods. Human designers are only able to solve complex problems by reducing the search space to a manageable size, through modular design for example. Evolution is able to search a larger solution space for solutions that better utilise the available resources, than could be found by a human designer. An extreme example of this is where evolution is able to utilise the analogue properties of a digital device.

### Autonomous Robotics Domain Highlighting EHW Capabilities

One specific application domain of EHW that has been particularly successful is autonomous robot control. Evolvable hardware controllers offer several advantages over classical control methods. Firstly, as a subset of the evolutionary robotics approach, it offers a solution to the difficult task of manually decomposing a control system, constructing behaviours, and ensuring that unpredictable interactions between behaviours don't occur. Secondly, it allows the full exploitation of the physical medium, disregarding constraints introduced by the conceptual model of the designer that prevent the characteristics of the hardware from being explored and exploited. Evolvable hardware is also able to utilise the inherent parallelism and asynchronous nature of hardware, rather than requiring these properties to be emulated on a sequential processor. Lastly, EHW offers the ability of adapting to electrical or mechanical faults; for example, if a robot leg is damaged, the controller could be reconfigured to adapt the robot gait to redistribute the robot's weight across the functional legs.

## 1.2   What is Evolvable Hardware

**Aim of Evolvable Hardware**

The aim of *evolvable hardware* (EHW) is to generate electronic circuits, using simulated evolution, that produce the behaviour specified by the designer. Rather than designing the circuit using traditional engineering methods, a fitness function is provided that tests each evolved circuit, and assigns it a measure of fitness according to how well it performs. After generations of mating circuits that have achieved some progress in producing the desired behaviour, the end result is a population containing one or more circuits that performs the desired task.

**Reconfigurable Hardware as Enabling Technology for EHW**

While the ideas behind EHW have been around for some time, it wasn't until reconfigurable hardware devices, such as field programmable gate arrays (FPGAs), became commercially available in the 1990's, that the field of EHW was able to become a reality.

Reconfigurable hardware devices, usually referred to as programmable logic chips, are made up of an arrangement of simple elements that have configurable connectivity between elements, and allow the simple elements to be configured to perform different functions. Programmable logic arrays (PLAs) are on the simple end of the scale, being comprised of an array of AND gates, with outputs connected to an array of OR gates. These allow an arbitrary combinatorial Boolean function, expressed as a *Sum of Products* to be performed. While on the complex end of the scale, a recently commercially available FPGA, the Virtex-4 (Xilinx Inc. 2004), provides a lattice of up to 200K configurable digital logic blocks (providing combinatorial and sequential functionality), specialised signal processing blocks, embedded processors, embedded RAM blocks, multiple dedicated clock routing lines, analog to digital converter, Gigabit transceiver, Ethernet MAC (Media Access Controller), system temperature and voltage monitoring, and many other features. Devices such as this allow the construction of an entire integrated hardware and software system on a single chip, giving rise to the term *System on a Chip* (SoC).

FPGAs are configured by downloading a bitstream to the device, which specifies the functions performed by the basic elements and the connectivity between elements, producing the desired circuit. In some FPGAs portions of the FPGA are able to be reconfigured on the fly, taking on the order of a few milliseconds. This has given rise to the field of reconfigurable computing, whereby portions of a hardware system can be brought on chip as required, allowing the system to provide more functionality than can be fit on the chip at one time.

In the context of EHW, FPGAs allow evolving circuits to be tested in situ, which is well suited to embedded applications such as autonomous robot controllers, image and signal processing.

## 1.3    The Main Problems to be solved in EHW

### Scalability

In his 2001 doctoral dissertion, Paul Layzell pointed out that, to his knowledge, "no circuit with 100+ functional basic elements has yet been evolved; the greatest number so far attained seems to be around 30-40" (Layzell 2001). Since this time, there has been little progress in scaling EHW to more complex problems.

Traditionally EHW has been done by manipulating the logic-gate level resources, such as multiplexors and Boolean function generators, on an FPGA. This has been done by encoding the device's configuration directly on the chromosome. This approach is known as a *direct encoding*, and is severely limited in its ability to scale to larger problems.

### High-level Primitives For Scaling EHW

An alternative to this is to evolve higher-level primitives such as adders, oscillators, or filters, etc. Although this scales the ability of EHW to solve more complex problems, it comes at the price of higher gate counts (Vassilev and Miller 2000), designer bias and loss of potential novelty in solutions, thus countering some of the original motivations for EHW. Also when using higher-level building blocks, even small genetic modifications may cause radical changes in behaviour, causing the evolutionary process to degenerate to a random walk (Layzell 2001).

### Morphogenesis For Scaling EHW

Another approach is to encode a growth process, known as *morphogenesis*, on the chromosome, rather than the actual FPGA configuration. This is how nature is able to evolve complex organisms. Rather than explicitly describing the resulting structure, the chromosome encodes a growth process which interacts with the current state of the growing structure, and through feedback, unfolds and guides the structure to its fully developed state.

This approach has increasingly been seen as a means of scaling EHW to more complex problems without losing its ability to generate novelty. However, how to encode and implement growth for EHW is still an open question, and to date there has been little success in applying morphogenesis to gate-level EHW on FPGAs.

4

**Difficulties with Complexity of Modern FPGAs**

Compounding this has been the increasing complexity of modern FPGAs. Commercially available FPGAs, that can be used for EHW, have complex logic blocks containing several look up tables (LUTs) that may be used for logic or RAM, several levels of signal routing, and may be set to potentially damaging configurations. This has resulted in a move away from utilising the properties and architecture of the underlying device. Instead, evolution is often performed on a virtual FPGA architecture, containing simple logic elements with unidirectional connections to neighbouring elements. By abstracting away the underlying architectural details EHW has moved away from one of its key advantages, of being able to discover novel solutions free of designer bias.

## 1.4   Motivation For This Research

At the commencement of the work presented in this book, the intention was to apply evolvable hardware to the task of creating cooperative reactive controllers for a hexapod walking robot with joint angle sensors and force-feedback on each leg. To develop a detailed design for independent cooperative controllers would be a difficult task, due to the complex nature of interactions between components in a high dimensioned phase space.

**EHW vs ANNs**

While artificial neural networks (ANNs) would appear to be the logical choice for solving such a problem, EHW has also successfully been used for solving control problems, and has comparable learning performance (Higuchi, Iwata, Kajitani, Yamada, Manderick, Hirao, Murakawa, Yoshizawa, and Furuya 1996; Torresen 2001). Though both EHW and ANNs can be implemented in hardware (the local cluster neural network (Koerner, Rueckert, and Sitte 1998) is one such example of a hardware neural network implementation) to provide superior performance, through massive parallelism, over microprocessor-based controllers, evolvable hardware offers several advantages.

Digital EHW is analysable in terms of Boolean functions (Higuchi, Iwata, Kajitani, Yamada, Manderick, Hirao, Murakawa, Yoshizawa, and Furuya 1996), and more significantly, EHW in general has been claimed to provide faster learning performance (Albert 1997; Higuchi, Iwata, Kajitani, Yamada, Manderick, Hirao, Murakawa, Yoshizawa, and Furuya 1996; Kajitani, Murakawa, Nishikawa, Yokoi, Kajihara, Iwata, Keymeulen, Sakanashi, and Higuchi 1999), with Kajitani (Kajitani, Murakawa, Nishikawa, Yokoi, Kajihara, Iwata, Keymeulen, Sakanashi, and Higuchi 1999) reporting an instance of EHW learning the same task as a backpropagation network (control of a prosthetic hand) in a thousandth of the time.

ANNs are also constrained by their architecture, with typically, only the weights being updatable. Logic gate-level EHW on the other hand, has few constraints; evolution is free to utilise all the properties of the device to come up with the best solution free from the designer's preconceptions. Although EANNs (Evolutionary ANNs, see (Yao 1999) for a survey) lack the restrictions of ANNs in general, they are implemented in software, and so cannot compete with the true massively parallel nature of EHW or hardware-based ANNs.

### Demise of EHW-Friendly FPGAs

Gate-level EHW based on FPGA technology seemed to offer great promise to applications in this area, however, with the demise of the EHW-friendly Xilinx 6200 series, FPGAs with complex architectures must now be used. The result of this is that even for simple applications, gate-level EHW is now forced to solve a much more complex problem.

### Need For Managing Mainstream FPGA Complexity

Faced with this problem, the original aims of developing EHW robot leg controllers had to be refined. Unless the complexity of gate-level EHW on modern mainstream FPGAs can be managed, EHW will be unable to solve the intended problem, or many other similar problems for which it was seen to be an approach with great promise.

To overcome this problem, the central focus of this work shifted from applying EHW to solving an autonomous robot control problem to solving the problem of how to scale EHW at the gate-level on a commercially available FPGA platform so that in future gate-level EHW can once again be successfully applied by researchers to a wide variety of difficult engineering problems using commercial-off-the-shelf hardware.

The aims of this project are covered in more detail in the following section.

## 1.5    Aim of the Project

### Central Aim

The central aim of this work is to scale EHW to increases in circuit size and complexity on *commercially available* reconfigurable hardware without requiring architectural details to be abstracted away. Although currently commercially available reconfigurable hardware devices (typically FPGAs) are not the most evolution-friendly platforms, having complex architectures and issues with potentially damaging configurations, evolving circuits on commercially available devices *without* requiring a move to high-level building blocks is a necessary prerequisite for the adoption of EHW to solving *real* problems in electronic design (in domains that are difficult for human design methods), repair and adaptation.

**Proposed Solution**

The proposed solution is a morphogenesis process that is closely tied to the reconfigurable hardware's architecture, and is supported by genetic representation and operators that allow this process to be encoded and manipulated by evolution in a manner that assists the search for circuits matching the desired criteria.

**Evaluation of Proposed Solution**

To determine if this approach is feasible, a gate-level *morphogenetic* EHW (MGEHW) model is developed for a Xilinx Virtex FPGA (a popular commercially available FPGA). This is then applied to solving various circuit generation problems, and its performance compared with a traditional direct encoding approach to EHW. Throughout this book, the number of generations required to find the desired solution is used as the performance indicator for both approaches to EHW.

**Proof of Concept**

It should be stressed that given the time limitations of the project, the work presented in this book is primarily a proof of concept, demonstrating that a morphogenetic approach to gate-level EHW on a mainstream FPGA is, firstly, viable, and secondly, can indeed provide a means of scaling EHW without requiring architectural details to be abstracted away.

# 1.6    Outline of Book

### Chapter 2: Evolvable Hardware Background

Chapter 2 covers the background to evolvable hardware necessary to understand the work presented later in the book. Evolvable hardware is defined, its components covered in detail, and the levels of hardware abstraction that are accessible to manipulation by evolution are identified. A literature survey of work done in EHW is presented, highlighting its applications and advantages. The various commercial and research reconfigurable hardware devices that have been (or may possibly be) used, for EHW are covered, and a commercially available device is chosen for the work conducted later in the book.

The problems faced by EHW are reviewed, along with proposed solutions to these. In particular, the scalability problem is identified as a major obstacle to progress in EHW, and morphogenesis proposed as a potential solution.

7

### Chapter 3: Morphogenesis Background

Chapter 3 investigates the properties of morphogenesis that may make it suitable for solving the scalability problem faced by EHW, and covers the biological background of morphogenesis, to identify the mechanisms that are essential to its functioning. Some existing models of morphogenesis from evolutionary computation and evolvable hardware are also covered to gain ideas as to how morphogenesis may best be implemented. With the knowledge garnered here, an effective morphogenetic model can then be designed and constructed.

### Chapter 4: Designing a Morphogenetic EHW System

In Chapter 4 a morphogenetic model for EHW is outlined based on the processes and mechanisms from Chapter 3 that can usefully be applied to EHW. The construction of this model involves creating a mapping between FPGA hardware resources and biological constructs, deciding on which developmental processes and mechanisms should be implemented to manipulate the hardware, and providing both a means of encoding the developmental process on the chromosome, and a set of genetic operators for evolution to manipulate this process specification.

### Chapter 5: Morphogenetic EHW Implementation

Chapter 5 provides details of the implementation of the gate-level morphogenetic EHW (MGEHW) system for the Xilinx Virtex FPGA based on the design given in Chapter 4. The system is broken up into the major components of its process flow, and each is described in detail.

### Chapter 6: Experiments with Evolving Circuit Structure

With a working morphogenetic EHW system in place, Chapter 6 proceeds to evaluate the performance of the morphogenetic EHW and traditional direct encoding EHW approaches to evolving (primarily) circuit structures. In particular the performance of these approaches in terms of scaling to increases in circuit size and complexity is measured.

### Chapter 7: Simulated Secondary Developmental Mechanisms

In Chapter 7 an investigation is conducted into whether adding simulated secondary EHW developmental mechanisms, to the purely hardware configuration state driven processes presented in earlier chapters, is able to further improve the performance of the morphogenetic approach. This chapter covers details of developmental mechanisms from biology that require simulated molecules, and their design and implementation for incorporation into the MGEHW system. Experiments are conducted to compare the performance of the MGEHW system, with and without these additions, for generating circuit structures.

## Chapter 8: Experiments with Evolving Circuit Functionality

Chapter 8 investigates the evolution of circuit structure and functionality on a more complete set of FPGA resources by both the morphogenetic (without the addition of simulated molecules) and direct encoding EHW approaches. At the same time, the effectiveness of various LUT encodings in aiding evolution to generate functional circuits is evaluated. Further experiments are carried out to isolate the cause of difficulties encountered by both EHW approaches in generating the desired circuit (a one-bit full adder).

## Chapter 9: Estimation of Problem Solvability

Chapter 9 analyses the results of the experiments presented in Chapters 6 and 8 to determine what factors determine whether a given EHW problem will be solvable (by the morphogenetic EHW approach in particular), and uses these to develop a heuristic for identifying whether or not any given problem is likely to be solvable by the morphogenetic EHW approach.

## Chapter 10: Conclusion

Chapter 10 concludes the book with a summary of the work presented in earlier chapters, and goes on to discuss issues and insights that have been discovered in the course of this work, along with avenues for future research.

## Appendix A: Exploratory Experiments

Appendix A contains the results of exploratory experiments with the morphogenetic EHW system that were used to determine appropriate system parameters prior to running the experiments in Chapters 6 and 7. These experiments were also used to gain insight into the working of the morphogenesis system and the evolution of chromosome structure during experimental runs. Comparative speed tests of direct encoding and morphogenetic approaches (with and without the addition of simulated molecules), and a sensitivity analysis of the parameters used by evolution and morphogenesis are also presented here.

## Appendix B: Experiment Complexity Calculations

Appendix B provides details of the equations and calculations (and the architectural and problem specific details required for these) that are used in Chapter 9. Calculations are provided for each experiment set, and summaries of the results are provided.

**Appendix C: Genetic Code Details**

Appendix C supplies the details of all the genetic codes used in the experiments conducted in this book. Details of the MGEHW system's genetic code, detailing how FPGA and simulated resources are encoded, and conversion (from an intermediate format) to Xilinx's JBits Java API format (for configuring and querying the Virtex FPGA's resources) are provided. The direct encoding scheme, used by the traditional EHW approach, is also provided for each experiment set.

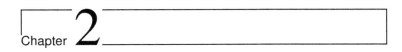

Chapter **2**

# Evolvable Hardware Background

The purpose of this chapter is to provide the necessary background for show-ing the value and strengths of evolvable hardware, while identifying its current shortcomings. The issue of scalability is identified, and an approach to solving this is introduced. Furthermore, possible platforms for conducting evolvable hardware research are investigated, and a suitable commercially available plat-form is chosen.

The first section defines evolvable hardware, and its components of simu-lated evolution and reconfigurable hardware are covered in some detail. Then, in Section 2.2, a literature survey is presented, showing the wide range of the task evolvable hardware has been applied to, and highlighting its advantages over conventional circuit design and software-based approaches, in an in-depth review of evolvable hardware work in robotics and online adaptation.

Circuit evolution is able to be performed at various levels of abstraction, or in other words, granularity of basic hardware components and their con-nections. This is covered in Section 2.3, along with a review of the hardware platforms that have been used in evolvable hardware. Of these, the Xilinx Virtex is one of the few commercially available platforms that can be used for evolvable hardware at a logic-gate level of abstraction. As such, this platform and its configuration is covered in detail in Section 2.4.

Then, in Section 2.5, evolvable hardware's current major limitation is cov-ered, along with a possible solution that is covered in detail in the following chapter, before concluding the chapter with a summary of the findings of this chapter.

## 2.1 Definition

Evolvable hardware (EHW) involves two aspects, that being simulated evolu-tion and electronic hardware, with which it aims to generate a circuit with a specific behaviour.

Evolvable hardware (EHW) can be defined as the automatic generation of electronic circuits using simulated evolution. Within this classification exist two main streams of work depending on whether the main focus is to evolve the circuit design, or the behaviours of the circuit. The former involves the application of evolutionary techniques to circuit synthesis, while in the latter view, which is that taken in this book, it is the behaviour of the circuit rather than the layout which is of primary importance. This is the case in circuits evolved for classification or reactive control, where generalisation ability is the aim. EHW is different from hardware implementations of evolutionary algorithms (EAs), where the hardware architecture doesn't change and is simply used to implement evolutionary functions such as selection, recombination and mutation.

**Reconfigurable Hardware**

EHW typically utilises reconfigurable hardware, usually programmable logic devices (PLDs) such as field programmable gate arrays (FPGAs), which allows evolving solutions to be tested in situ, and as such is well suited to embedded applications such as robot controllers and image processing.

Field programmable logic arrays consist of a lattice of cells, with the outer cells connecting to pins on the chip for input and output, and the inner cells being either memory blocks or configurable logic blocks (CLBs). CLBs typically consist of some arrangement of basic logic gates, multiplexors, and flip-flops that can be configured to perform various digital logic functions. Figures 2.3 and 2.4, in Section 2.4, show the structure of a Virtex FPGA and (half of) a logic block, as representative of FPGAs in general. The functionality of the cells and the connections between cells can be configured to produce the desired circuit. The architecture and function of an FPGA is determined by a set of architecture bits that can be reconfigured.

When FPGAs are used in EHW, they are coupled with simulated evolution, which is used to evolve a good FPGA configuration in order to solve a particular problem. The fitness of a candidate solution is evaluated according to the behaviour of the configured FPGA circuit.

**Simulated Evolution**

Simulated evolution has its roots in evolutionary computation, which is covered next, in Section 2.1.1. A popular classification of EHW is based on whether the evolved circuits are evaluated in hardware or simulation, hence this is covered in Section 2.1.2.

## 2.1.1    Evolutionary Computation Overview

The field of Evolutionary Computation (EC) takes its inspiration from nature, specifically from the Darwinian theory of evolution with its emphasis on the survival of the fittest, and applies it to search and optimisation problems. Here, it is candidate solutions to the problem at hand that must compete for survival and the opportunity to reproduce, and through the mating of fit candidates and the chance of opportunistic mutations, honed by natural selection, drive evolution to produce increasingly fit solutions.

All approaches to evolutionary computation generally start with a randomly seeded population of candidate solutions, to which selective pressures are applied, such that reproduction is biased towards individuals with some degree of fitness. After some number of generations the evolutionary process will, hopefully, have converged on a population containing at least one optimal or near-optimal solution, and many good solutions. The general scheme for an evolutionary algorithm is shown in Figure 2.1.

Each member of the population is comprised of a set of parameters, or genes. The set of specific value of the genes in the individual (the genotype) determine the properties (phenotype) of the candidate solution. Candidates are assessed for their suitability and awarded a fitness score. Evolution is guided through the fittest candidates reproducing more than the less fit, or in other words, through natural selection. Simply reproducing the fittest candidates is not enough, some means of finding new, hopefully better candidates is required (i.e. the search space must be extended to new areas). This is achieved through the mutation and combination (or crossover) operators during reproduction.

### Evolutionary Operators

Crossover acts as a means of attempting to create better candidates through the combination of, hopefully good, parts of the genotypes of fit candidates. This allows the evolutionary process to move towards promising regions of the search space. Mutation occurs during reproduction by randomly adding small amount of noise to the genotype, typically making a small change to a gene in the genotype of the offspring. Mutation is used to prevent premature convergence to local optima by randomly sampling new points in the search space.

### Applications of EC

Evolutionary computation has been extensively used for solving optimisation and other intractably complex problems where no easy method of human decomposition is available to allow a sufficiently accurate model to be applied. On the other hand, due to the stochastic nature of evolutionary computation,

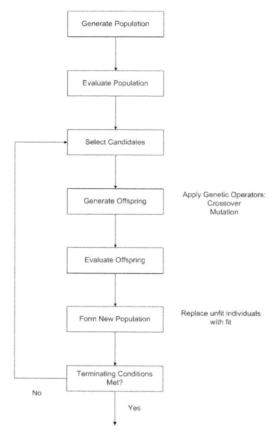

Figure 2.1: Outline of Evolutionary Algorithm

there is no guarantee that the algorithms will converge (on the optimal or near-optimal solution).

Being a zeroth order method (it requires only values of the function to optimise), evolutionary computation is able to be applied to optimisation problems where standard optimisation techniques, such as gradient descent that requires the existence and computation of derivatives, are not applicable (Schoenauer and Michalewicz 1997). This enables evolutionary computation approaches to find the global optimum of very rough functions, unlike traditional methods that are local in scope and so would tend to only find the local optimum closest to their starting point.

## Main Evolutionary Computation Approaches

There are four main approaches in evolutionary computation. These are Genetic Algorithms (GA), Genetic Programming (GP), Evolutionary Strategies (ES), and Evolutionary Programming (EP).

## Genetic Algorithms

Genetic algorithms were originally developed for finding optimal solutions to sequential decision processes (Jong 1992), but have tended to be used for performing function optimisation. GAs have typically used bit-strings for encoding the genome, although more recently other encodings have been utilised according to the problem at hand. Selection scheme has traditionally been proportional selection, where the probability of an individual being selected for reproduction is proportional to its fitness, although ranking selection (based on the individual's rank in the population) and tournament selection (choosing the best individual from a uniformly chosen subset) have become more popular choices. In early GAs, all parents are replaced by their offspring, which may cause the best fitness in the population to decrease, and for this reason, more recent work favours the use of elitist replacement schemes, where the top individuals are kept.

In GAs, both crossover and mutation are applied for generating offspring. Crossover works by replacing a segment of the genome of one parent, with the corresponding segment of the other. Mutation applies random bit flips to the offspring's genotype with a preset probability.

The most significant difference between GAs and the other EC approaches is that GAs manipulate the genotype, with a mapping from genotype to phenotype being used to evaluate the fitness of the genotype. The other approaches operate directly on the phenotype.

## Genetic Programming

Genetic programming originated as an application of GAs to tree-like structures for evolving computer programs (Koza 1992). Through the use of Lisp-like S-expressions, it is easy to define a closed crossover operator. That is, by swapping sub-trees of two valid S-expressions, the resulting S-expression can be ensured to be valid. However, in the original GPs there was no mutation operator. More recently, mutation operators have been introduced, such as random replacement of a subtree, node, or leaf. A steady state replacement scheme is usually used, where the offspring of parents chosen by tournament

15

is placed in the population and then the individual to be removed is that with the lowest fitness taken from a uniformly chosen set.

### Evolutionary Strategies

Evolutionary strategies originated as an algorithm for dealing with parameter optimisation problems, and in its original form manipulated a single-member population comprised of a real valued vector. In this approach the individual is operated on solely using Gaussian mutation, which involves adding a zero-mean Gaussian variable of standard deviation (s), to create a new candidate. The fittest out of the old and new candidates is kept, for further mutation, while the less fit is disregarded.

The main operator for ES is mutation and is typically, in the usual case of a population of real-valued vectors, self-adaptive mutation, where the standard deviation parameter of Gaussian mutation is included in the genotype and subject to the evolutionary process. As evolution progresses, the mutation rate (standard deviation) is reduced adaptively through the evolution process itself. This adaptive mutation rate aids evolution in converging on a good set of solutions. Adaptation of learning rates during the optimisation process is also utilised in other techniques such as gradient descent, Kohonen feature maps, and learning vector quantisation (LVQ), where it may be useful or necessary to ensure convergence. Aside from mutation, ES often uses a global recombination operator that involves all the individuals in the population.

### Evolutionary Programming

Evolutionary programming was initially developed for constructing finite state machines (FSMs) which arose from the classical AI problem of predicting the next state in a series of environmental stimuli, represented as symbols from a finite alphabet. EP works on a population of FSMs, which are evaluated by comparing the outputs of the FSM to the next input signal and assigning a fitness based on their predictive performance.

Like early ES, EP only uses a mutation operator, and as in ES every individual in the population generates offspring, in this case one per parent. EP uses an elitist replacement scheme; the best N individuals in the population (of parents and offspring) are kept. However, in recent advances, a stochastic tournament replacement scheme may be used, other solution spaces may be handled including self-adaptive Gaussian deviations for real-value variables (Schoenauer and Michalewicz 1997). The generalisation of evolutionary programming to continuous value optimisation makes it almost equivalent to evolutionary strategies.

## 2.1.2    Intrinsic vs. Extrinsic

Approaches to evolvable hardware can be classed as intrinsic or extrinsic according to the manner in which the population of candidate solutions is evaluated.

**Extrinsic EHW**

In extrinsic EHW evolution of the hardware is done in simulation and when the evolutionary process has found a suitable design, the fittest configuration is then implemented in hardware. Alternatively, the best configuration in each generation may be downloaded onto the hardware if a reconfigurable device is used.

**Intrinsic EHW**

Intrinsic EHW evaluates the fitness of each phenotype directly in hardware. In other words, each chromosome in the population is decoded and downloaded onto the reconfigurable hardware device. Hence, an intrinsic approach is able to take advantage of all the dynamic parameters within the device and the real-world conditions encountered during evolution, whereas extrinsic EHW is evolved to the constraints of the mathematical model in which it is simulated.

On the other hand, unconstrained intrinsic evolution may adapt solutions too tightly to the physical properties of the device. Thompson's early work on evolving a tone discriminator on a Xilinx 6200 series FPGA (Thompson 1996b) was a good example of this; however, in his later work (Thompson and Layzell 2000) it was shown that this needn't be the case. Experiments showed that intrinsically evolved solutions were able to work correctly under a wide range of conditions, even being transferable to other hardware devices of the same type. This counters most of the supposed advantages of extrinsic evolution, these being that it is able to offer potentially greater reliability and reproducibility in the evolved circuit (Albert 1997).

Other advantages of intrinsic EHW are speed, scalability and accuracy. Intrinsic evolution can speed up a search for a solution by several orders of magnitude compared with evolution in software, especially for large complex analogue circuits, and scales well with both circuit size and model accuracy (Stoica, Keymeulen, Tawel, Salazar-Lazaro, and Li 1999).

**Deciding between Intrinsic and Extrinsic**

Possibly the most important factor in deciding between intrinsic and extrinsic EHW, for non-online adaptive EHW (i.e. EHW not used for in situ fault tolerance and adaptation) is the extent of required accuracy. For the evolution of combinatorial logic, or purely theoretical work, this is not an issue. However, for circuits intended for real-world applications, a solution evolved in software may behave differently when downloaded in programmable hardware, due to the limited accuracy models of physical hardware used by software simulators. Circuits containing sequential logic are liable to have differences between simulated and actual behaviour, as they may have their behaviour affected by variable gate delays due to environmental conditions (temperature for example), and changes in signal ordering. This problem is avoided if the circuit is evolved directly in hardware, or if only synchronous signals are used (i.e. through the use of flip flops).

## 2.2  Applications

EHW has been applied to various problems, including evolving combinatorial logic (Miller, Job, and Vassilev 2000; Gordon and Bentley 2005; Stomeo, Kalganova, Lambert, Lipnitsakya, and Yatskevich 2005), sequential circuits (Garvie and Thompson 2003; Shanthi, Singaram, and Parthasarathi 2005), feature extraction (Porter 2001), image processing (Hollingworth, Tyrell, and Smith 1999; Torresen 2000b), digital filters (Miller 1999; Tufte and Haddow 2000; Sekanina and Růžička 2003; Vinger and Torresen 2003), analog filters (Lohn and Colombano 1999), tone discriminators (Thompson 1996b), amplifiers (Lohn and Colombano 1999), oscillators (Huelsbergen, Rietman, and Slous 1999; Aggarwal 2003), data compression (Higuchi, Iwata, Keymeulen, Sakanashi, Murakawa, Kajitani, Takahashi, Toda, Salami, Kajihara, and Otsu 1999), adaptive equalization (Higuchi, Murakawa, Iwata, Kajitani, Liu, and Salami 1997), digital to analog converters (Zebulum, Keymeulen, Duong, Ferguson, and Stoica 2003; Langeheine, Meier, Schemmel, and Trefzer 2004), circuit repair (Moreno, Madrenas, Cabestany, Canto, Kielbik, Faura, and Insenser 1999; Vigander 2001), fault tolerance (Keymeulen, Zebulem, Jin, and Stoica 2000; Miller and Hartmann 2001; Prodan, Tempesti, Mange, and Stauffer 2001; Stoica, Keymeulen, and Zebulum 2001; Garvie and Thompson 2004; Liu, Miller, and Tyrrell 2005b), adaptive sensors (Keymeulen, Zebulem, Stoica, and Buehler 2001), analog controllers (Zebulum, Pacheco, Vellasco, and Sinohara 2000; Gwaltney and Ferguson 2005), DC motor control (Gwaltney and Ferguson 2003), prosthetic control (Kajitani, Murakawa, Nishikawa, Yokoi, Kajihara, Iwata, Keymeulen, Sakanashi, and Higuchi 1999), autonomous robot navigation (Keymeulen, Konaka, Iwata, Kuniyoshi, and Higuchi 1998) and control (Thompson 1995; Haddow and Tufte 1999; Berenson, Estévez, and Lipson 2005).

Unlike conventional hardware, EHW is able to adapt online to faults or changes in environment. This is one of its greatest strengths over traditional electronic design paradigms. EHW also has many benefits that make it attractive to the creation of robot controllers. This is particularly the case for complex controllers required to work in real time, such as cooperative reactive controllers for a hexapod robot with force feedback, which was the initial motivation for the work presented in this book. Hence, while there have been many areas to which EHW has been applied, as illustrated above, the application of EHW to online adaptation and robot control will be concentrated on this section.

## 2.2.1    Online Adaptation

Current approaches usually evolve a static circuit design, optimised for certain situations, but unable to change behaviour once evolved. However, EHW has the capacity, unlike conventional hardware, to be able to use the evolutionary process to adapt to changes in the environment or in the task requirement by reconfiguring its hardware structure dynamically and autonomously.

### Electrotechnical Laboratory's Adaptive EHW Chip

Researchers in the Electrotechnical Laboratory, Japan, developed an adaptive evolvable hardware chip for adaptive control which incorporates the evolutionary process on the same chip as the evolving design (Higuchi, Iwata, Takahashi, Kasai, Sakanashi, Murakawa, and Kajitani 2000). The population of candidate chromosomes is stored on chip, as is the genetic algorithm, and two programmable logic arrays in which two candidate circuits may be tested in parallel. Thus, adaptation can be achieved through evolution. This EHW chip was used to implement a prosthetic hand controller (Kajitani, Murakawa, Nishikawa, Yokoi, Kajihara, Iwata, Keymeulen, Sakanashi, and Higuchi 1999) and for learning to navigate an autonomous wheeled robot, for tracking a coloured ball while avoiding obstacles and adapting to sensor failures (Higuchi, Iwata, Keymeulen, Sakanashi, Murakawa, Kajitani, Takahashi, Toda, Salami, Kajihara, and Otsu 1999).

The performance of this online adaptive EHW controller was shown to be remarkable in terms of speed and degree of adaptation in comparison with other autonomous robots. The robot was generally able to find within 10 minutes, on average, a set of suitable hardware configurations for the reactive robot controller, and was able to switch in real-time between configurations to track the ball while avoiding obstacles. The authors reported that this was two orders of magnitude faster than previous work.

In the same work, the authors showed that the adaptive EHW controller was, within a few minutes, able to autonomously find a new set of hardware configurations that allowed the robot to continue tracking the ball when some of the robot's sensors were blinded.

19

### Complete Hardware Evolution Approach

Another example of online adaptation in EHW is the Complete Hardware Evolution (CHE) approach of (Haddow and Tufte 2000). Here the evolution process can be triggered by changes in the hardware or environment to restart evolution if the task is no longer being solved satisfactorily by the current FPGA configuration.

In this approach, the genetic algorithm is implemented on the same chip (FPGA) as the evolving design, through the use of their GA pipeline. Feedback of fitness to the GA pipeline allows adaptive evolution. This design was used to develop their GERC robot controller, which was based on Thompson's dynamic state machine (DSM) work (see (Thompson 1995)). Their idea is that adaptation doesn't need to be decided at design time, instead, evolution could be restarted when the current design is no longer able to perform satisfactorily.

### FIPSOC, Embryonics and Other

A FIPSOC (Field Programmable System on a Chip) architecture was developed by (Moreno, Cabestany, Madrenas, Cantó, Faura, and Insenser 1999) capable of online adaptive EHW. Although there haven't been any demonstrations of evolving hardware solutions on it, it was shown to be able to perform self-repair through online reconfiguration (Moreno, Madrenas, Cabestany, Canto, Kielbik, Faura, and Insenser 1999).

Embryonics (Ortega-Sánchez and Tyrell 1997; Ortega and Tyrell 1997; Ortega-Sánchez and Tyrell 1998; Ortega and Tyrell 1998) and Immunotronics (Bradley, Ortega-Sanchez, and Tyrell 2000; Bradley and Tyrell 2000a; Bradley and Tyrell 2000b; Bradley and Tyrell 2001) also have some limited adaptation capability, although here there is no adaptation of behaviour, only of the configuration in response to failures in the device's cells. An implementation of an embryonic architecture on a Xilinx Virtex FPGA is given in (Ortega and Tyrell 2000).

In (Milne 1999) a model for self-reconfigurable autonomous adaptive hardware was presented, however, no EHW implementation was demonstrated.

### Robotics Application

An example of where online adaptation might be useful in EHW was given in (Layzell 2001). An EHW controller for a legged robot may be able to change its gait if a leg joint seized, right itself if placed upside-down, or find a way of walking over or around a previously unencountered obstacle. Another area of application would be in space exploration, as presented in (Darrin, Conde, Chern, Luers, Jurczyk, and Mills 2001) where an online adaptable hardware module is being developed for space instrument controllers (see also

(Vigander 2001) for an evolutionary approach to fault repair of circuits in space applications).

### Necessity of Online Adaptation

However, it is not certain that online adaptation is necessary for EHW in most cases. In the first case, as behavioural adaptation, it would be hoped that a suitably evolved design may be able to generalise to unforseen circumstances, as is achieved in neural network approaches. This was shown to be the case in (Keymeulen, Konaka, Iwata, Kuniyoshi, and Higuchi 1998) where an evolved controller was able to generalise to situations that it hadn't encountered in evolution.

In the second case, robustness of the evolved circuits to component faults and failures has been shown to be able to be achieved without explicitly needing to be pre-designed, as in embryonics and immunotronics, through the evolutionary process (Thompson 1996a; Thompson, Harvey, and Husbands 1996; Tyrrell, Hollingworth, and Smith 2001).

### Comparison of Population Evolution and Online Adaptation

A comparison between the fault tolerance of a population-evolution approach, as in online adaptive EHW where a population is kept and if a fault occurs another member may be used, and a fitness-evolution approach, where fault tolerance is tested for during evolution resulting in an individual that has this property, was made in (Keymeulen, Zebulem, Jin, and Stoica 2000). It was found that the fitness-based approach has the advantage of obtaining a single circuit resilient to multiple faults, however the population approach is able to find circuits in the population with better performance and requires 20% to 50% less computation than the fitness-based approach. Also, although the population approach offers a self-repair mechanism, it must be done off-line in 50-80% of the time needed to obtain a solution from scratch.

## 2.2.2    EHW Robot Controllers

### Classical Robotics Approach

Classical approaches to robotics decomposed the robot control problem into perception, planning and action modules, where the planning module received and fused the sensory information from the perceptual module to create a world model which was then used for planning a set of actions to be performed. This approach suffered from many problems, such as computational bottlenecks, noisy sensors, and the difficulty of modelling the complexities of reality, leading researchers to look for alternatives that would work in real time and be robust.

### Advantage of EHW Controllers

Evolvable hardware controllers offer several advantages. Firstly, as a subset of the evolutionary robotics approach, it offers a solution to the difficult task of manually decomposing a control system, constructing behaviours, and ensuring that unpredictable interactions between behaviours don't occur. Secondly, it allows the full exploitation of the physical medium, disregarding constraints introduced by the conceptual model of the designer that prevent the characteristics of the hardware from being explored and exploited. Lastly, evolvable hardware is able to utilise the inherent parallelism of hardware, rather than this to be emulated on a sequential processor.

### Early Work in EHW Controllers

An early work in the field of EHW was George Friedman's investigation into the automatic synthesis of circuits in his 1956 master's thesis (Friedman 1956). Friedman used the principles of variation and selection to develop a Selective Feedback Computer that designs control circuits for a simple autonomous mobile robot. This is done by varying the parameters and interconnections of basic neuron-like threshold circuits and selecting them according to their fitness in generating robot behaviours. The desired behaviour here being that a robot situated in a variable temperature field maintains its body at a constant temperature.

### Dynamic State Machine

In more recent times, Adrian Thompson applied EHW to a robot control problem with his work on the Dynamic State Machine (DSM) architecture (Thompson 1995), which, unlike FSMs, doesn't use a global clock to synchronize signals. This allows the controller to take advantage of the real-time dynamics of the sensory-motor environment, and to interface both analogue and digital components requiring both synchronous and asynchronous control (Haddow and Tufte 1999). Using this scheme, Thompson evolved a wall-avoidance behaviour on a hardware controller for a 2-wheeled robot autonomous robot with a pair of time-of-flight sonar sensors.

### Gate-level EHW Welding Controller

Higuchi (Higuchi, Iwata, Kajitani, Yamada, Manderick, Hirao, Murakawa, Yoshizawa, and Furuya 1996) used gate-level evolution to develop a V-shape ditch tracer as a part of a welding robot controller. When two steel components are welded, a V-shaped ditch is formed along which a welding robot waves its arm. This requires the robot to be able to detect the centre of the ditch. Two sensors on the robot arm measure the distances from the V-shape, and by comparing these a two-channel comparator can detect whether the arm is in the centre or not. An EHW controller was able to learn and synthesise a comparator circuit on-line and to take over from the existing comparator.

### Evolutionary Autonomous Robot Navigation

(Keymeulen, Durantez, Konaka, Kuniyoshi, and Higuchi 1996; Keymeulen, Konaka, Iwata, Kuniyoshi, and Higuchi 1998; Higuchi, Iwata, Keymeulen, Sakanashi, Murakawa, Kajitani, Takahashi, Toda, Salami, Kajihara, and Otsu 1999) did work on an evolutionary autonomous robot navigation system which combines learning and evolution: they evolve a gate-level hardware configuration for a reactive Boolean-function controller in a simulator and then after reaching the target 64 times they download the controller of the fittest individual onto the FPGA robot controller which is able to reach the target in an unseen environment.

### Hexapod Robot Controller

An EHW controller for a hexapod robot was evolved in the work of Johnson, et al. (Johnson, Parker, Cyliax, and Braun 1997), but unlike the approach in (Thompson 1995) where the hardware controller is evolved directly, here a finite-state machine implementation is evolved which is translated into a hardware description language and from which a net list is synthesised for configuring the FPGA controller. In this approach the authors used a Cyclic Genetic Algorithm (CGA), an adaptation of basic GAs for developing cycles of sequential instructions, to represent the gait controller, while the rest of the controller is statically defined and not affected by evolution. Also, in this approach, although the complexity is higher in terms of actuators (12 versus the 2 motors in Thompson's robot) there is no sensor feedback to the controller.

### Genetically Evolved Robot Controller

Another example is Haddow and Tufte's work on Complete Hardware Evolution (CHE) and their experimental robot controller design, Genetically Evolved Robot Controller (GERC) (Haddow and Tufte 1999). They used Lego's Mindstorm for the robot body and motors, with two light detectors in front and sensors for detecting wheel rotation, and based their controller on Thompson's DSM Architecture (see above). A controller was evolved for making the robot go as far and as straight as possible, in the test environment, while avoiding walls in its path.

**Prosthetic Hand Controller**

An EHW prosthetic hand controller providing six different motions in three different degrees of freedom was developed for adapting the controller for dedicated users (Kajitani, Murakawa, Nishikawa, Yokoi, Kajihara, Iwata, Keymeulen, Sakanashi, and Higuchi 1999; Torresen 2001; Higuchi, Iwata, Keymeulen, Sakanashi, Murakawa, Kajitani, Takahashi, Toda, Salami, Kajihara, and Otsu 1999). The EHW controller showed similar performance to an artificial neural network (ANN), and fit within smaller hardware.

### 2.2.3    Advantages of EHW Controllers

#### Advantage Over Software and Sequential Processors

Evolvable hardware controllers offer several advantages over software-based approaches. EHW allows the physical medium of the control system to be utilized, removing arbitrary constraints that prevent the characteristics of the hardware from being explored and exploited. Importantly, EHW is able to utilise the inherent parallelism and asynchronous nature of hardware rather than requiring these properties to be emulated on a basically sequential processor.

EHW is often implemented on FPGAs, which have lower power consumption and heat dissipation than an equivalent processor due to their considerably lower clock rates: an FPGA running at a much lower MHz can give comparable performance to a sequential processor.

#### EHW Performance Comparable to Artificial Neural Networks

Adaptive controllers are often implemented as artificial neural networks, and these require a floating point CPU or a neural network chip.

Torresen (Torresen 2001) showed that using gate-level EHW provides similar performance as an artificial neural network, with a much more compact hardware implementation, which makes it more feasible to embed the controller (in a prosthetic hand in this case).

Higuchi et al. (Higuchi, Iwata, Kajitani, Yamada, Manderick, Hirao, Murakawa, Yoshizawa, and Furuya 1996) demonstrated that EHW could be used as an alternative to ANNs in classification and control tasks. They were able to successfully develop a welder controller, pattern recognition circuits, a 4-state automaton, and circuits that could perform comparably with ANNs on the classification of Iris data, transformations of 2D images (rotation), and distinguishing two intertwined spirals.

## EHW Faster Than ANNs

Some authors have claimed that EHW is faster than ANNs (Albert 1997; Higuchi, Iwata, Kajitani, Yamada, Manderick, Hirao, Murakawa, Yoshizawa, and Furuya 1996). This claim was supported by work done on a EHW chip for controlling a prosthetic hand (Kajitani, Murakawa, Nishikawa, Yokoi, Kajihara, Iwata, Keymeulen, Sakanashi, and Higuchi 1999) where the authors made a direct comparison to an ANN controller using back-propagation. Off-line learning on the ANN took around three hours, compared to less than a second for the EHW chip. The EHW solution was also slightly better in terms of learning, achieving an 85% average action rate compared to an 80% average for the ANN.

## Evolved Design Analysable

Another advantage of (digital) EHW over ANNs, as noted in (Higuchi, Iwata, Kajitani, Yamada, Manderick, Hirao, Murakawa, Yoshizawa, and Furuya 1996) is that the evolved circuit can be readily analysed in terms of Boolean functions. A variable length encoding, would help in the analysis as it would be likely to eliminate non-participatory elements from a solution, which were, in direct encoding approaches, basically noise due to the fixed length of a bit-stream encoding. In (Thompson 1996b) Thompson showed how to analyse an evolved circuit to determine which parts are participating.

However, to allow logic analysis of an evolved circuit requires disallowing unconstrained evolution, which means disallowing evolution from utilising the analogue properties of a digital chip. One way of achieving this is the approach of (Levi and Guccione 1999b; Levi and Guccione 1999a) where asynchronous set and resets and transparent latch mode on the flip-flops are disallowed.

## EHW as Alternative to Conventional Control Circuit Design

A comparison between conventional control circuit design, involving finding a Laplace transfer function for the compensator and then finding a circuit that implements this transfer function, and evolvable hardware was made in (Zebulum, Pacheco, Vellasco, and Sinohara 2000). The EHW approach was seen to have advantages in that no intermediate Laplace transfer function needs to be found for the comparator, no human knowledge derived from design techniques needs to be supplied, and rather than sampling standard circuit topologies, novel implementations may also be investigated which may result in circuits with advantages in terms of the number of components, and noise, dissipation and fault tolerance aspects.

## 2.3   Abstraction Level and Platform

As mentioned earlier, evolvable hardware requires two things, these being simulated evolution and reconfigurable hardware. In EHW, evolution is applied to the construction of circuits by configuring and connecting components on a reconfigurable device in various manners. These components can range from elementary transistors all the way to complex components, such as filters for example. This defines the level of abstraction, and accordingly a reconfigurable hardware platform that supports this level is required.

The most popular reconfigurable devices are field programmable gate arrays, which allow evolution at the logic-gate or higher level, while there are analogue platforms which allow evolution at lower levels of abstraction. Some non-commercial research platforms have also been developed which have other basic components. These levels of abstraction and platforms are covered in the following sub-sections.

### 2.3.1   Abstraction Level

**Gate-level EHW**

Gate-level evolution manipulates the architecture of the reconfigurable logic device (typically an FPGA) at the level of the configurable multiplexors (logic gates), Boolean function generating look-up-tables (LUTs), registers and random access memory blocks (BRAM). While, strictly speaking, FPGAs are not configurable at the level of *basic* logic gates (such as NANDs), the manipulation of the FPGA's lowest level of configurable components (multiplexors, LUTs, registers and RAM) by simulated evolution is known as gate-level EHW.

This approach, which traditionally directly modified the configuration bit stream, has enabled evolution to come up with circuit designs free from designer bias, at times even utilising the analog and physical properties of the digital device as in Thompson's work on unconstrained evolution (Thompson 1996b).

**Function-level EHW**

An alternative to this approach is to use higher-level primitives, high-level hardware functions, such as adders, sine generators, etc (Higuchi, Murakawa, Iwata, Kajitani, Liu, and Salami 1997). This approach is known as function-level EHW. This approach aids evolution in generating more complex circuits, at the price of introducing designer bias, and increasing the number of gates required (Vassilev and Miller 2000). Examples of this approach include (Higuchi, Murakawa, Iwata, Kajitani, Liu, and Salami 1997; Clark 1999; Kalganova 2000b).

**Analogue EHW**

EHW has also been applied to generating analogue circuits using lower-level components (than digital logic gates) such as operational amplifiers , transistors, and other analogue components. Analogue work was initially done extrinsically using Spice simulations ((Koza, Andre, Bennet III, and Keane 1997) and (Lohn and Colombano 1999; Lohn, Haith, Comombano, and Stassinopoulos 2000) are examples of this), however more recently work has been able to be done using intrinsic methods due to the development of reconfigurable analogue devices such as Field Programmable Analog Arrays (FPAAs) (Anderson 1998; Flockton and Sheehan 1998; Ozsvald 1998) and Field Programmable Transistor Arrays (FPTAs) (Stoica, Keymeulen, Tawel, Salazar-Lazaro, and Li 1999; Keymeulen, Zebulem, Jin, and Stoica 2000).

**Analogue Advantages**

With this has come a renewed interest in applying evolutionary techniques to analogue devices. Analogue circuitry has advantages in terms of costs, size and power, and can directly process signals that are continual in time and amplitude, albeit at a finite resolution (Stoica, Keymeulen, Tawel, Salazar-Lazaro, and Li 1999).

A single transistor is able to perform many functions that can be used in computation, such as the generation of square, square-root, exponential and logarithmic functions. Transistors can also act as voltage-controlled current sources, long- or short-term analogue storage, and perform analogue multiplication of voltage (Stoica, Keymeulen, Tawel, Salazar-Lazaro, and Li 1999).

There is also a current belief that evolutionary search works better with analogue, rather than digital, circuits (Stoica, Keymeulen, Tawel, Salazar-Lazaro, and Li 1999) possibly due to the fact that analogue behaviours provide relatively smoother search spaces.

**Analogue Hinderance**

The main hinderance to performing EHW on reconfigurable analogue devices at this time is the lack of commercially available platforms suitable for EHW. This is elaborated further in Section 2.3.3.

## 2.3.2    Field Programmable Gate Arrays

Field programmable logic arrays consist of a lattice of blocks, with an outer ring of IO blocks (IOBs) that connect to the IO pins on the chip, and inner blocks that are comprised largely of configurable logic blocks (CLBs), with the remaining blocks being dedicated memory blocks (BRAM). CLBs typically consist of some arrangement of basic logic gates, multiplexors, and flip-flops that can be configured to perform various digital logic functions. The functionality of the various blocks and the connections between them can be configured to produce the desired circuit.

### Xilinx 6200

The release of the Xilinx XC6200 series of fine grained FPGAs was largely responsible for the explosion of work in EHW in the mid- to late- 1990's. The 6200 series had an architecture which had many evolution friendly features, such as an open bitstream with partial, dynamic reconfiguration at the bit level in which all configurations are legal, and a cell interconnect strategy that prevented harmful configurations from occurring. Besides Adrian Thompson's ground breaking work using the 6200 (Thompson 1996b), there was much other work including (Fogarty, Miller, and Thomson 1997; de Garis and Korkin 2002; Huelsbergen, Rietman, and Slous 1999; Wee, Park, and Lee 1999).

In the XC6200 family of FPGAs, each cell (fine grained CLB) receives 3 inputs (X1,X2,X3) to its function unit. This function unit is able to act as a 2:1 multiplexor, or perform any Boolean function of 2 variables, chosen from the cell's inputs. Each cell receives a signal from each of the neighbouring cells (N, S, E, W), and from the four directional FastLANES (N4, S4, E4, W4), that are routed across the 4x4 block of cells. In turn, each cell is able to output its function unit's output to its 4 neighbours (N, S, E, W), while cells on the edge of a 4x4 (or 16x16 or 64x64) block of cells are also able to route a signal to the neighbouring block using the Magic wires (depending on the function implemented by the cell, however) (Xilinx Inc. 1997).

An XC6200 cell is illustrated in Figure 2.2, with the input (X1-X3), output (Nout, Sout, Eout, Wout) and Magic multiplexors shown, and the function unit demarked, which includes the Y2 and Y3 input select and inversion, RP (register protect), and CS (combinatorial/sequential) output select multiplexors. From this it can be seen that the cell performs its Boolean function on the X2 and X3 input signals (alternatively, feedback from the cell's register may be used), as determined by the Y2 and Y3 multiplexors and the X1 input signal. Also, each cell output, bound for a neighbouring cell, may either simply route a signal from one of the other cell's neighbours, or provide the output of the function unit. Note that the Magic and RP multiplexors were generally not manipulated in EHW, nor were IOBs and FastLANEs (the N4, S4, E4, W4 inputs).

Unfortunately, while the 6200 series was a boon to researchers, its fine-grained cells (with 13 multiplexor elements) and limited modes of interconnection meant that it was unsuitable for most commercial applications. Thus it was phased out of production.

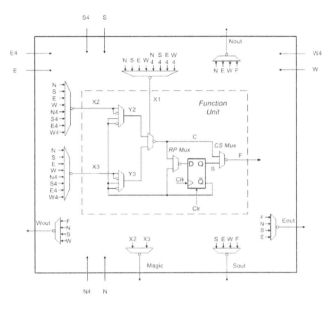

Figure 2.2: Xilinx XC6200 Cell

## Xilinx 4000

The Xilinx 4000 series, with a coarser grained CLB (that can handle up to
9 inputs and has 2 flip flops) was never particularly attractive for research in
EHW, lacking partial reconfiguration. (Levi and Guccione 1999b) was one of
the few who reported doing EHW work using the 4000 series.

## Xilinx Virtex

With the release of the Xilinx Virtex series, the Virtex has become the digital EHW platform most currently used. It offers partial reconfiguration, and
although the Virtex architecture is course grained (each CLB can handle functions with up to 19 inputs and has 4 flip flops), fine grained access is possible
through the JBits Java API (Guccione, Levi, and Sundararajan 1999).

Much EHW work on the Virtex, however, uses a virtual 6200 architecture
for which circuits are evolved and then mapped onto the Virtex. This is mainly
to prevent illegal configurations that may damage the device. See (Haddow
and Tufte 2001) and (Hollingworth, Smith, and Tyrell 2000) as examples of

29

this.

Xilinx has also released successors to the Virtex, such as the Virtex-E, Virtex-II (with optional embedded processor cores), and the Virtex-4, of these it is only the Virtex-2 which has fine-grained access provided, through JBits version 2.

Although other FPGA manufacturers exist, they have attracted little attention in EHW.

### 2.3.3    Analogue Platforms

**Field Programmable Analog Arrays**

Field programmable analog arrays (FPAAs) are the analogue counterpart of FPGAs, likewise consisting of a lattice of configurable blocks with programmable interconnections. Here, however, the blocks are nominally identical, typically comprised of an operational amplifier (op amp), and known as configurable analog blocks (CABs).

Whereas FPGAs were developed for applications in reconfigurable computing and digital signal processing, FPAAs were typically developed for applications in programmable mixed-signal circuits and filtering (Keymeulen, Zebulem, Jin, and Stoica 2000).

With the demise of the Motorola *MPAA020* field programmable analog array, prior to the commencement of this work, there was only one commercially available FPAA (which has since also become obsolete), the Zetex *TRAC020* chip, which consists of 20 configurable op amps as basic elements. Although there has been EHW work reported using this device (Flockton and Sheehan 1998; Ozsvald 1998), Flockton and Sheehan (Flockton and Sheehan 1998) noted that the Zetex TRAC device doesn't have any on-chip programmable resistors or capacitors, and so needs external components provided to produce useful circuits. In an update to their previous work, Flockton and Sheehan combined the Zetex device (op amps) with the Evolvable Motherboard (see (Layzell 1999; Layzell 2001)) for its resistors, to evolve bandpass filters (Flockton and Sheehan 2000).

Lattice Semiconductor produces a simple FPAA, the *ispPAC30* (In-System Programmable Analog Circuit), aimed at amplification and filtering applications (Ramsden 2001). It is comprised of of 4 differential inputs, 4 programmable input amplifiers, 2 output amplifiers, 2 multiplying digital-analog converters, programmable interconnects between these, and 2 internal adjustable voltage references. However, it uses electrically erasable configuration memory, which limits the number of times it can be reconfigured (to somewhere in the order of 10,000 times).

Anadigm provides a family of FPAAs, comprised of a matrix of fully configurable switched capacitor cells, known as configurable analog blocks (CABs), embedded within a fabric of programmable interconnect resources. CABs are comprised of an op amp, capacitor banks, and local routing, switching and clocking resources, as well as global connection points. Configuration is performed by downloading a bitstream to the chip, and the 4x5 CAB *AN10E40* device is reprogrammable in 125 microseconds (Anadigm Inc. 2002). These devices are coarse grained, and configured using Anadigm's design software library, making them generally unsuitable for evolvable hardware (although they have been used to create artificial neural networks which may have their parameters evolved, as was done in (Berenson, Estévez, and Lipson 2005)).

Besides being limited by the number of reconfigurable elements (20 appears to be the upper limit), FPAAs also have problems in terms of bandwidth, switch resistances, accuracy and noise. See (Zebulum, Stoica, and Keymeulen 2000) for an outline of work being done to address the current limitations of FPAAs.

(Ramsden, Greenwood, and Hunter 2005) have recently developed an alternative analog platform to commercially available FPAAs, known as the Evolvable Analog Research Platform (*EARP-1*). This platform provides course-grained functional blocks, similar to those on the ispPAC30, suitable for supporting a signal-flow circuit model, but with more interconnect flexibility than commercially available FPAAs. Furthermore, simple IO and programmability are provided. At this time no EHW work has been reported using this platform.

### Field Programmable Transistor Arrays

More recently field programmable transistor arrays (FPTAs) have been developed at NASA's Jet Propulsion Laboratory, offering a finer granularity than previous EHW platforms, being programmable at the transistor level (Stoica, Keymeulen, Tawel, Salazar-Lazaro, and Li 1999). According to (Keymeulen, Zebulem, Jin, and Stoica 2000) this feature is advantageous from an EHW perspective, allowing novel architectures to be sampled and standard architectures to be implemented. This lets the designer set the level of granularity, so that the evolutionary process may manipulate the cell at the transistor level, or freeze the cells architecture to high level analogue or digital building blocks for evolution to manipulate (Zebulum, Stoica, and Keymeulen 2000).

The FPTA cell is an array consisting of 8 transistors interconnected by 24 programmable switches implemented with transistors, acting as simple T-gate switches. The status of the switches (ON or OFF) determines the circuit function and topology. The FPTA architecture allows the implementation of larger circuits through cascading FPTA cells. To offer sufficient flexibility, all transistor terminals are connected to expansion terminals via switches.

The original chip the PTA (Stoica, Keymeulen, Tawel, Salazar-Lazaro, and Li 1999) had 3 cells inter-connectable only through external wiring. This was succeeded by the FPTA1 which provided a proper FPTA with 12 cells and included capacitors. The most recent model is the FPTA2 which consists of an 8x8 matrix of reconfigurable cells, and is able to receive 96 analogue/digital inputs and provide 64 analogue/digital outputs. Each cell interconnects with its north, east, south, west neighbours. An array of 16x8 photodetectors is integrated on the chip, being distributed within the cells (Stoica, Zebulum, and Keymeulen 2001).

The FPTA2 chip has been integrated into a stand-alone board-level evolvable system (SABLES) (Stoica, Zebulum, Ferguson, Keymeulen, and Duong 2002). SABLES is comprised of the FPTA2 chip connected via a 7.5Mhz 32-bit bus to a TI DSP that implements the evolutionary algorithm which controls the reconfiguration of the FPTA.

Another group has developed a prototype FPTA chip (Langeheine, Foelling, Meier, and Schemmel 2000; Langeheine, Foelling, Meier, and Schemmel 2001) consisting of an array of 16 x 16 programmable transistor cells, with each cell containing a configurable array of 4 x 5 (giving 20) transistors providing 5 different lengths and 4 different widths. PMOS and NMOS transistor cells are placed in alternation. The authors state that with 256 transistors in total, circuits of fairly high complexity should be evolvable, and have attempted intrinsic evolution of digital-to-analog converters with voltage mode output and 6-bit resolution (Langeheine, Meier, Schemmel, and Trefzer 2004).

Although FPTAs have much to offer, they are not currently available on the market, limiting their use to those with the facilities to design and fabricate them.

### 2.3.4   Other Architectures

**Evolvable Motherboard**

A general purpose evolvable testbed, the Evolvable Motherboard (Layzell 1999; Layzell 2001), was developed by Paul Layzell as an alternative to FPGAs (and FPAAs) with the primary objectives being to provide a much larger range of basic configurable elements, to allow virtually any interconnection between these elements, and to allow access to individual elements by test equipment for analysis. It is comprised of a diagonal matrix of analogue switches, connected to up to 6 plug-in daughter-boards, which contain the basic electronic components (anything from resistors to microprocessors) for evolution. However, Layzell stresses that the evolvable hardware motherboard is not designed for producing large scale applications; instead it is intended to function as a research tool for investigating different evolvable architectures, algorithms, and the circuits they produce on a small scale (Layzell 1999).

### Processing Integrated Grid

A cellular-automata (CA) like architecture, the Processing Integrated Grid (PIG) architecture, is given in (Macias 1999a; Macias 1999b). The PIG platform is an example of a fine grained self-reconfigurable massively parallel architecture. Each cell receives inputs from its four neighbours, and uses its internal truth table to provide the four outputs which go to its neighbours. Cells may also modify their neighbour's truth tables, allowing both data and functions to be transferred through the grid.

### CAM-Brain

Another CA-like architecture, the CAM-Brain, is a multi-FPGA (Xilinx XC6264) based 3D CA based neural network that uses a GA to evolve the neural modules (de Garis and Korkin 2002). It has been developed as a research tool for creating artificial brains.

### Field Programmable System on a Chip

A FIPSOC (Field Programmable System on a Chip) architecture was presented in (Moreno, Cabestany, Madrenas, Cantó, Faura, and Insenser 1999; Moreno, Madrenas, Cabestany, Canto, Kielbik, Faura, and Insenser 1999) which was composed of a programmable logic section (like an FPGA) with fast dynamic reconfiguration, a configurable analogue section, and a microcontroller. It was developed for applications in online adaptive EHW, online learning for ANNs, and self-repairing hardware, such as might be useful in space applications.

### POEtic Project

The POEtic project has developed a specialised bio-inspired neuromorphic hardware system comprised of a 2D array of functional units that have the ability to evolve, self-repair and grow, and learn (Tyrell, Sanchez, Floreano, Tempesti, Mange, Moreno, Rosenberg, and Villa 2003). The POEtic tissue is organised into three layers (Tempesti, Roggen, Sanchez, and Thoma 2002): a genotype layer that contains the genetic information for the organism, a configuration layer that implements cellular differentiation and growth, and a phenotype layer that implements learning using spiking neurons.

## 2.4   The Xilinx Virtex FPGA

Currently the mainstream (commercially available) platform most used for EHW is the Xilinx Virtex. It offers partial reconfiguration, and fine grained access through the JBits Java API (Guccione, Levi, and Sundararajan 1999).

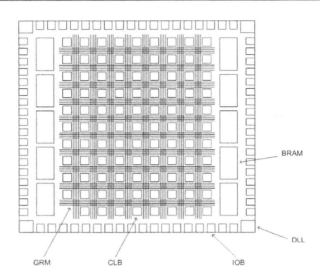

**Figure** 2.3: Virtex Structure

In this section, the Virtex architecture is covered after which various means of configuring the device are discussed, of which JBits is the only viable means for gate-level EHW, and as such, is covered in detail.

### 2.4.1    Virtex Architecture

The Virtex provides CLBs, block ram (BRAM), input output blocks (IOBs) and several layers of routing between them via the general routing matrix (GRM), including single lines to the neighbouring four blocks, hexlines to blocks six blocks away, longlines that span a row or column, and horizontal tristate busses between cells. The Virtex also provides dedicated clock signal routing with low skew (Xilinx Inc. 2001), and dedicated delay-locked loops (DLLs) for clock-distribution delay compensation and clock control. The region adjacent to the IOBs (not labelled) contains additional routing resources for improving I/O routing. The general structure of a Virtex FPGA is shown in Figure 2.3.

Each CLB is divided into two slices, each consisting of two lookup tables (LUTs) which may be used as 4-input (1-output) function generators, 16-bit shift registers, or as 16x1-bit synchronous RAM. Associated with each LUT is an output flip flop (configurable to positive or negative edge triggered and may have synchronous or asynchronous reset) to create a basic logic element.

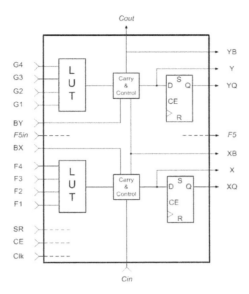

Figure 2.4: Virtex Slice Structure

The general structure of a Virtex slice is shown in Figure 2.4. Slice LUTs may also be combined to provide functions of five or six inputs, or 16x2-bit or 32x1-bit RAM. The CLB also has dedicated carry in and out lines, with associated carry propagation logic, which is useful for multi-bit arithmetic. Each slice can implement a 2-bit adder and by stacking CLBs, flexible bit-width arithmetic with low propagation delays can be implemented. A dedicated AND gate is also provided in the carry logic, with which integer multiplication can be efficiently implement, effectively halving the number of CLBs required. IOBs and BRAM are also configurable, but as these are not generally manipulated by EHW systems, they are not covered in detail here.

General signal routing between CLBs, IOBs and BRAM, is done via the general routing matrix, for which there are switch boxes adjacent to CLBs (also the BRAM and IOBs) through which horizontal and vertical routing resources connect. There are several resources available via the GRM, including single-length lines (singles) that connect the switch matrix to the adjacent GRMs in each direction, and hex lines that are used to connect to a CLB 6 blocks away. Half of the single lines (24 in each direction) from one CLB's output bus are directly connectable to the adjacent CLB inputs.

## 2.4.2    Virtex Configuration

The Virtex device is configured by downloading a configuration bitstream to it. It allows partial reconfiguration, by which CLBs can be independently addressed in frames. There are 48 frames associated with each column and a number of rows of CLBs per column (depending on the actual device). To reconfigure a particular CLB, multiple frames are required, each of which reconfigures a frame of the CLB column.

The details of this bitstream is proprietary and largely secret and thus can't be manipulated directly to the extent required for EHW (see (Carmichael 1999) for details on the open portions of the bitstream). Generally, however, the configuration bitstream is generated by FPGA design software from schematics or a hardware description language such as VHDL. The level of abstraction provided by FPGA design software doesn't allow low-level access to the specifics of the particulars of the FPGA structure (such as configuring routing wires or inner-CLB multiplexors). Also, to go from a hardware design specification to an FPGA configuration requires a few intervening steps, such as compilation and implementation (place and route) phases, which can be time consuming. Hence, this approach is unattractive for work in EHW.

FPGA design software, however, generates a placed and routed FPGA data file, in NCD format, which is then fed through the Xilinx bitgen.exe to create the configuration bitstream. The NCD format is open, and there is a low-level ASCII equivalent, XDL, and tools to translate between the two formats, hence it is possible to configure the device manually using an XDL specification (Seaman 2000), although the amount of documentation available on the XDL format is limited. The alternative to these approaches, and the one generally used by the EHW community for evolving circuits on Xilinx Virtex series FPGAs is JBits.

## 2.4.3    Configuration with JBits

Xilinx provides fine grained access to the Virtex FPGA structure through the JBits Java application programming interface (API) (Guccione, Levi, and Sundararajan 1999). JBits allows access to all the configurable resources at the multiplexor (gate) level. This includes look up tables (LUTs) that implement Boolean functions, flip flops, routing lines, etc., which are able to be individually configured to create or modify circuits on the FPGA. JBits operates on a configuration bitstream which is then downloaded onto the FPGA ready for use.

In Figure 2.5 the gate-level internal slice logic is shown. The resources in the diagram that are available for JBits to manipulate are the LUT contents (16 bits), LUT functionality configuration logic (labeled as *Write Strobe Logic* and *Data In Mux Logic*), register modes (flip flop or latch, synchronous or asynchronous reset) and internal configuration muxes (the unmarked muxes). JBits also provides access to the routing muxes that configure the general routing matrix (GRM), tristate bus, local to CLB lines and CLB to CLB directly connecting lines, illustrated in Figure 2.6.

JBits can be used both for configuring and querying the FPGA logic-gate (mux) level configuration, as required for implementing an EHW morphogenesis process driven by the gate level state.

Resources are accessed using the *set()* and *get()* methods on the FPGA bitstream, indexed by CLB or BRAM row and column (or IOB side and index), while the *bits* parameter specifies the particular resource to configure or query. The *set()* method also has a *val* parameter that allows the resource to be configured to one of a number of predefined settings.

The *bits* and *val* constants for a given resource (multiplexor) are defined in an associated resource class within the *com.Xilinx.JBits.Virtex.Bits* package. For example, the *OUT0* class defines the mux associated with bit zero in the CLB output bus. Within this class a *bits* constant, with the same name as the class, is defined to represent this resource, for example, *OUT0.OUT0*. This particular mux selects one (or none) of the CLBs outputs to drive on line 0 of the CLB out bus. Hence, for each of the available CLB outputs to select there is an associated *val* const. For example *S0_YQ* represents the registered output of slice 0's G LUT.

Using this example, a call of *get(row, col, OUT0.OUT0)* will return the current multixplexor configuration as one of the constants defined within the *OUT0* class. A returned value of *OUT0.OFF* would indicate that no CLB output has been selected to drive this bus line. To reconfigure this line to accept input from *S0_YQ*, a call of *set(row, col, OUT0.S0_YQ)* would be made.

# 2.5    Problems in EHW

There are two main problems encountered with gate-level EHW on FPGAs, the first of these, ensuring the validity of generated configurations, is, in a sense, incidental but nevertheless important, while the second, scalability, is major and is a limiting factor on work in this area. Furthermore, scalability is an issue in the evolution of circuits on any platform, and more so its bearing covers the whole realm of evolved structures. These two issues are covered in detail below.

## 2.5.1    Circuit Validity Issues

The appearance of the Xilinx 6200 family of FPGAs was largely responsible for making experimentation in the field of EHW accessible to a larger research community. This FPGA family offered many evolution friendly features, in-

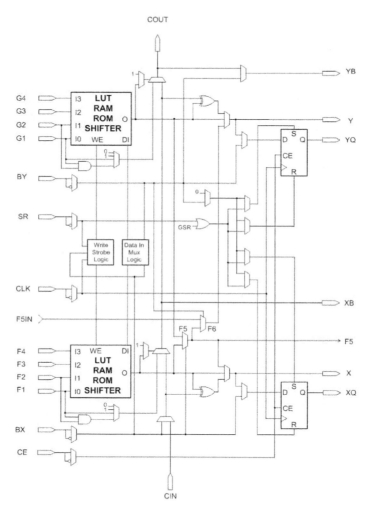

Figure 2.5: Virtex Slice Muxes

38

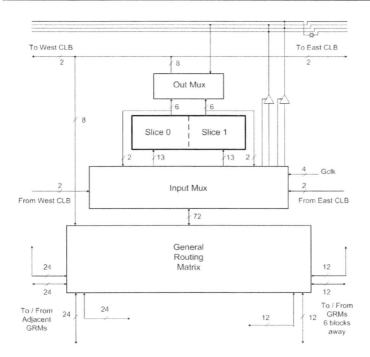

Figure 2.6: Virtex CLB Routing

cluding, an open architecture in which all configurations were valid, and partial reconfiguration down to a single bit.

However, since the demise of the 6200 series, there have been no readily available EHW platforms in which evolved configurations can be guaranteed to be valid, in the sense that they won't result in circuits that may damage the device. On the other hand, by not directly manipulating the bitstream and instead using a suitable vendor-supplied configuration interface, such as Xilinx's JBits API for the Virtex, the configuration bitstream itself can be ensured to be valid.

## Contention

Invalid, or damaging, configurations can occur through contention on the wires, whereby more than one output is driving the same wire with opposing voltages resulting in high currents (Levi and Guccione 1999a). (See also (Hadžić,

Udani, and Smith 1999) for a discussion on how damaging configurations can be deliberately realised in the context of FPGA viruses).

Recently, the Xilinx Virtex series of FPGAs has become the most popular of the mainstream digital EHW platforms. Unlike the 6200 series, the Virtex architecture is course grained, but fine grained access is possible through the JBits Java API (Guccione, Levi, and Sundararajan 1999). Routing representation in the Virtex is an issue, as each CLB is driven independently. This means that it is possible for contention to arise between two drivers if output multiplexors from different CLBs drive the same line.

Thus, the validity of evolved solutions has become an issue for modern EHW platform.

### Rerouting

The most obvious approach requires examining the configuration bitstream before downloading to the EHW device, and if contention is detected then the wires are rerouted until there is no more contention (Levi and Guccione 1999a). This approach, although valid, can be extremely computationally expensive, thus slowing the evolutionary process.

### Prevention by Construction

The approach taken in the work of (Levi and Guccione 1999b; Levi and Guccione 1999a) is to use a Java-based software tool, GeneticFPGA, that uses a pluggable mapper and translate modules to map the chromosomes bit stream into a FPGA circuit configuration, for Xilinx XC4000 EX/XL devices. To prevent illegal configurations from occurring, the authors used a method of prevention by construction, whereby genes are only able to manipulate uncontentious resources. Wires and resources which have more than a single driver, and thus able to cause contention, are simply turned off, and because they have no genetic representation they are unable to evolve connections to the circuit (Levi and Guccione 1999a).

### Virtual CLBs

Hollingorth, Smith and Tyrrell (Hollingworth, Smith, and Tyrell 2000) approached the problem of safe intrinsic evolution on Virtex devices by using an array of XC6200-like CLBs within the device, with each virtual CLB fitting into two Virtex CLBs. Here, evolution acts on the configuration of the virtual CLBs, rather than arbitrarily on the device's configuration.

Haddow and Tufte recently presented an EHW-friendly virtual architecture based roughly on the Xilinx 6200 family (Haddow and Tufte 2001). This architecture was developed to allow reuse of results and lessons in EHW through the use of a virtual platform which can be mapped onto any given technology.

From this architecture there is a simple mapping onto a Virtex architecture, with two virtual CLBs, or 'sblocks', being mapped into a single Virtex CLB, with only connections between CLBs and some of the CLB internal routing being used. This ensures that all genotypes result in legal phenotypes on the Virtex.

### Removal of Contentious Lines

Whereas previous approaches to intrinsic evolution on Virtex devices used fixed routing (Hollingworth, Smith, and Tyrell 2000) for contention avoidance, the contention avoidance method taken by Bentley et al. (Bentley, Gordon, Kim, and Kumar 2001) was to modify the representation. The authors noted that although CLB input multiplexors can connect to many nearest neighbour routing lines (singles), the connections to these lines is sparse for output multiplexors, with few that can connect to any that their neighbours can. The authors found that it was only necessary to prohibit eight of the possible forty-eight (directly connecting) singles. Only singles were encoded on the chromosome, and connection points between singles weren't evolved.

## 2.5.2    Scalability

### Direct Encoding

Current approaches to EHW generally use a direct encoding, where the chromosome is actually the device's configuration bit stream (Thompson 1996b), a connectivity matrix (Higuchi, Iba, and Manderic 1994) or node specification array (Thompson, Harvey, and Husbands 1996).

In all these approaches, the functions of each node and the connections between nodes (in a connectivity matrix), or the sources of each node (in the node specification approach), are directly encoded onto the chromosome. An example of a direct encoding for the Virtex is illustrated in Figure 2.7.

### Problems with Direct Encodings

As (Kitano 1996b) pointed out, such an encoding scheme has serious drawbacks in terms of the speed of convergence, i.e. the time required to find a solution. As the search space grows exponentially with the size of the problem, search algorithms that use a direct encoding will not scale up to larger problems (Hornby, Lipson, and Pollack 2001).

(Thompson 1996b) reported that using a direct encoding at gate-level required 1800 bits to configure a 10x10 corner of a Xilinx XC6216 FPGA. For more recent FPGA families with more complex CLB structures, interconnection strategies, IO and other details, then the number of configuration bits is likely to be several orders of magnitude more. As an example, the Xilinx

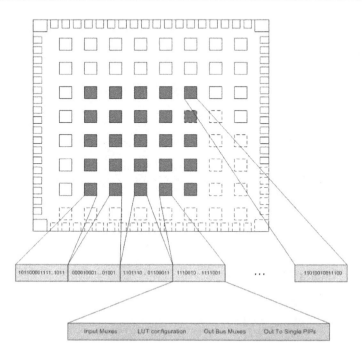

Figure 2.7: Direct Encoding EHW

V1000 with a CLB matrix of 40x32 requires around 6 million configuration bits (Carmichael 1999). (Yao and Higuchi 1999) reported that it was not unusual to take days to evolve a circuit with just 100 functional components.

Other limitations of the direct encoding approach include the size and overhead of the chromosome, and possibilities of illegal configurations. The latter wasn't an issue with early work in EHW, which generally used the Xilinx CX6200 family of FPGAs in which all configurations were legal. It may be an issue with other FPGA families; however the Xilinx JBits interface (Guccione, Levi, and Sundararajan 1999) takes care of issues of illegal configuration in some of the more popular Xilinx FPGA families currently in use.

### Scalability Limitation

The most important limitation, however, is in scaling EHW to more complex problems. Function-level evolution has approached the problem by using high-level hardware functions (such as addition, subtraction, sine, etc.) as primitives, instead of simple logic functions. See (Higuchi, Murakawa, Iwata, Kajitani, Liu, and Salami 1997; Clark 1999; Kalganova 2000b) as examples of this approach.

### Approaches Addressing Scalability

Increased complexity evolution (Torresen 1998) has approached the problem of evolving complex systems through a hierarchical approach of evolving components first, then their interconnections after. This involves a decomposition of the problem such that solutions to subsets of the task are evolved first, and then later combined by hand. However, the problems with which this approach was developed require subsets that aren't interdependent, thus allowing an easy decomposition. This method was applied to evolving an autonomous steering control system on a Xilinx XC6200 FPGA (Torresen 2000a).

Kalganova's bidirectional incremental evolution (Kalganova 2000a) incorporates 2 main approaches: divide-and-conquer and incremental evolution. The principle is to divide a complex task into simpler sub-tasks which are evolved and then merged incrementally. This approach was used to evolve a 7-input 10-output logic function. However, it appears that this approach is limited to evolving designs where both the inputs and outputs are already known, and also may be further limited to working only on logic functions due to their approach to decomposition. Kalganova's work on function-level EHW (Kalganova 2000b) also seems to be limited to similar domains.

More recently a similar approach was presented by Hong and Cho, called modular evolvable hardware (Hong and Cho 2003). It is aimed for use with digital circuits where there are clearly defined input/output patters and the number of outputs is fixed. Each module is defined as producing one of the circuit's outputs, and these are evolved separately, but with the evolved modules being reused in evolving other modules The modules (each of which has been evolved multiple times) are then combined by searching all combinations of evolved modules (combination being limited to a subset of each module's set of evolved solutions), and choosing the most optimised structure. This system was used to evolve 1- and 2- bit adders, and a 2-bit multiplier. According to the authors their method worked 50-1000 times faster than the conventional EHW approach, and further, produced more a optimised result. This work has similar limitations to that of Kalganova's approach (Kalganova 2000a).

### Encoding a Growth Process

In (Haddow and Tufte 2001), the authors state that it is becoming generally accepted that there needs to be a move away from a direct, or one-to-one genotype-phenotype, mapping to enable evolution of large complex circuits.

The alternative to a direct encoding which specifies the device configuration is to encode a growth process, called morphogenesis, thus moving the complexity from the genotype to the genotype-phenotype mapping.

## 2.6   Summary

In this chapter, the basics of evolvable hardware have been covered. Early work in EHW has shown the ability of evolution in generating novel circuits that exhibit superior performance over traditional approaches to circuit design, by utilising the properties of the underlying medium of the reconfigurable hardware device. Most importantly, EHW has shown its ability in adapting to dynamic environments and faults internal or external to the system.

To take advantage of the ability of EHW to utilise its underlying medium, the work in this book has chosen to focus on evolving circuits at a relatively low level, that being the logic-gate level (in terms of FPGAs, this effectively means the configurable multiplexors and look up tables, implementing Boolean function generators). To support this level of evolution, the Xilinx Virtex FPGA, accessed via the JBits API, is the only suitable commercially available platform.

However, evolution at the gate-level introduces two problems, the first of which is incidental, while the second is of primary importance. The first problem is ensuring that the FPGA configuration is valid, which primarily means in the context of the Virtex accessed via JBits, non-damaging to the hardware. The second problem is the scalability of EHW at the gate-level. This problem leads to the core issue of this book: how can gate-level EHW be scaled on a mainstream commercial FPGA?

Existing research has suggested that the solution to this problem is to encode a growth process on the evolved chromosomes, rather than directly encoding the FPGA configuration. This approach is known as morphogenesis, and is the topic of the next chapter.

44

# Chapter 3

# Morphogenesis Background

From the previous chapter, it can be seen that scalability is a major concern for evolvable hardware. In particular, due to the increases in FPGA complexity direct encoding approaches are no longer viable. Furthermore, if evolvable hardware is to be applied to large, real-world design problems, then the issue of scalability must be resolved.

Morphogenesis has been proposed as an alternative to directly encoding the FPGA configuration on the chromosome (Haddow and Tufte 2001). In this chapter the properties of morphogenesis that may make it suitable for EHW are investigated in Section 3.1. Then, Section 3.2 covers the biological background of morphogenesis, so that the mechanisms that are essential to its functioning can be identified, allowing a model of morphogenesis to be constructed for EHW. Following this, in Section 3.3, some existing models of morphogenesis from evolutionary computation (EC) and evolvable hardware (EHW) are covered. With this knowledge, an efficient morphogenetic model will be constructed in the following chapters.

## 3.1   Properties of Morphogenesis

In nature, the developmental process, which is referred to here as morpho-genesis, is neither externally defined nor explicitly represented in the genes. Instead, highly indirect chains of interacting 'rules' together specify the growth process for generating complex organisms. The flow of activation is not com-pletely predetermined and pre-programmed; it is dynamic, parallel and adap-tive. Specified through highly indirect chains of interacting rules, that are ap-plied iteratively and in parallel. Each gene's functionality is determined by its content on a variable length chromosome, and is only expressed when its pro-moter site is activated by proteins in the cell. Genes produce proteins, which in turn trigger the expression of genes, as well as controlling cell growth, dif-ferentiation, etc, thus creating networks of gene regulation or expression. The genotype acts as a constraint on the way in which an organism develops from a single cell such that symmetries and repetitions are able to emerge naturally in organisms (Thompson, Harvey, and Husbands 1996).

This approach has several properties that make it attractive to EHW, foremost amongst these is its ability to achieve greater scalability than other approaches (Hornby and Pollack 2001a), and through the encoding of repetition and symmetry (Thompson, Harvey, and Husbands 1996) evolve more complex phenotypes (Bentley and Kumar 1999) while representing the genotype more compactly (Hornby and Pollack 2001a) (than a direct encoding).

By separating the genotype (the genes encoded on the chromosome) space and the phenotype (the resulting structure or organism) spaces, and providing a mapping between them, in this case via a morphogenesis process, this gives a search space comprised of genotypes and a solution space of phenotypes. A genotype represents a point in search space and is operated on by genetic operators. A phenotype represents a point in solution space and, as with all evolutionary approaches, is evaluated by the fitness function (Banzhaf 1994).

This highlights another attractive feature of morphogenetic approaches, that being robustness to genetic operators and the ability to ensure the legality of phenotypes through the separation of search and solution spaces. Evolutionary search is further aided by the fact that a many-to-one mapping between genotype and phenotype allows movements in genotype space without requiring immediate changes in fitness, maintaining population diversity and encouraging exploration.

One of the foundations of biological evolution's success is its incremental nature, based on the property of modularity provided by the morphogenesis process. Modularity allows parts of the phenotype to be modified without interfering with the functionality of other components.

These properties of morphogenesis are also attractive to improving the performance of EHW. The following sections examine these properties in more detail.

## 3.1.1   Scalability of Morphogenesis

In traditional evolutionary computation genes are viewed as parameter settings on a fixed length genome at set loci. For evolving small or simple systems, it is adequate to directly encode the components' types, their parameters, and interconnections sequentially in the genotype. However, as a system becomes larger and more complex, then it can be expected that repetitions and symmetries are likely to play a more significant role.

An example of this, as given in (Thompson, Harvey, and Husbands 1996), is in the design of a control system for a robot with bilateral symmetry. In a case like this there would be good reason to expect the symmetrical nature of the robot to be reflected to some extent in the controller. However, if the genotype constitutes in effect a sequential description of each component of the system in turn, with the left and right halves separately described, then bilateral symmetry can only be achieved by separate independent evolution of each half of the genotype towards the same goal. In a stochastic process such as evolution this is improbable and difficult. However, it could be achieved relatively simply if the genotype described just one half, together with a routine that processes the description twice with appropriate parameters. In cases of even higher repetition, such as a retinal array, this becomes even more efficient (Thompson, Harvey, and Husbands 1996).

Morphogenesis has properties such as inherent parallelism of gene activation, multiple functions can be specified with the same genes, and genes may be activated and suppressed several times during development (Bentley and Kumar 1999). Here, rules are applied iteratively and in parallel, allowing the natural emergence of conditionals, implicit looping, recursion, and other advanced capabilities that must be manually added to other artificial development schemes, such as genetic programming.

To compare the performance of different genetic encodings, Bentley and Kumar conducted experiments on evolving tessellating tiles (Bentley and Kumar 1999). It was found that the morphogenetic approach (which they call an intrinsic embryogeny) constantly outperformed approaches using directly encoded GA, externally defined genotype-phenotype mapping, and GP program trees. The morphogenetic approach displayed remarkable efficiency and scalability, being able to evolve perfect solutions in fewer generations for every tile size, and without increasing the length of the genotype (the number of rules needed to specify the tile) which was in stark contrast to the other approaches.

## 3.1.2    Genetic Robustness and Phenotype Legality

Another attractive feature of morphogenetic approaches is their robustness of their encoding to genetic operators. This is in stark contrast to directly encoded GAs and genetic programming. In the former, even small mutations can produce drastic changes, and crossover has been often seen as damaging except early in an evolutionary run. In the latter, the creation of suitable representations that can be successfully evolved can be difficult, and often require the use of specialised genetic operators to minimise disruption (Bentley and Kumar 1999).

Morphogenetic approaches also provide advantages when evolving solutions for problems with constraints. In constrained optimisation problems a candidate solution is judged not just according to its fitness, but also according to how it fits the given constraints. In obeying constraints, entire regions of solution space may become unfeasible. Therefore in constrained optimisation, both the quality and feasibility of a solution have to be satisfied though these may be antagonistic. Often constraining the solution space leads to local hills or valleys which are difficult to overcome with traditional methods of optimisation (Banzhaf 1994). However, by separating the search and solution spaces (genotype and phenotype) and introducing a mapping between them the problem is greatly simplified. In the same manner that the result of evaluation gives the fitness of the phenotype, and by implication, of the underlying genotype; it is the phenotypes which must satisfy the problem constraints, not the genotypes (Yu and Bentley 1998). Hence, with separated genotypic and phenotypic spaces, unrestricted search operators can be applied to the search space and feasibility can be guaranteed in the solution space (Banzhaf 1994).

Other evolutionary approaches have often constrained or modified the genetic operators to aid the search process in finding, or retaining, suitable solutions; typically to avoid destroying good schema through recombination. Some examples include the SEAM (Symbiotic Evolutionary Adaptation Model) approach of Watson and Pollack (Watson and Pollack 2001; Watson and Pollack 2000), homologous crossover (recombination at points where chromosomes are most similar) (Burke, De Jong, Grefenstette, Ramsey, and Wu 1998), and the use of templates in (Jacob 1996) for controlling the scopes of operators. See (Yu and Bentley 1998) for a complete discussion on methods for ensuring legal solutions.

### 3.1.3    Aiding Evolutionary Search with Neutral Pathways

Separating the genotype and phenotype, as is done in a morphogenetic approach, also provides the opportunity to utilise many-to-one mappings, in which a number of different genotypes produce the same phenotype. Hence, it is possible to perform genetic operations, such as mutation, which modify the genotype without changing the phenotype. Such mutations are called neutral mutations (Shipman, Shakleton, and Harvey 2000). In Kimura's neutrality theory of evolution (Kimura 1983), evolution at the molecular level is theorised to be mainly due to mutations that are nearly neutral with respect to natural selection. Mutation and the resulting random drift of genomes are thus considered the main forces behind evolution. The majority of all mutations are neutral with only a minute fraction of non-neutral mutations being beneficial.

If evolution is viewed as a process of hill climbing on a fitness landscape, then when a population reaches a local peak, then for the population to be able to find a higher point, it is necessary to sacrifice its current fitness, temporarily accepting a lower fitness in the hope that a higher point can later be found. Although it is possible that the population could occupy several peaks and that recombination between individuals on different peaks could allow further progression, this relies on maintaining sufficient diversity in an evolving population, which goes against the pressures of selection, and random genetic drift (see (Harvey and Thompson 1996)). Neutrality in the genotype-phenotype mapping gives another possibility for overcoming this problem.

Neutral mutations allow movements in genotype space with no changes in fitness, which gives the population the ability to take mutations that are not immediately beneficial. Thus the population is able to drift along neutral ridges, rather than sacrificing its current fitness, which may significantly aid the evolutionary search. Instead of becoming trapped in sub-optimal regions of the landscape a population is able to continue moving through genotype space in search of areas that allow further increases in fitness.

This was shown to significantly aid evolutionary search for evolving telecommunications networks (Shipman, Shakleton, and Harvey 2000) and evolvable hardware (Harvey and Thompson 1996). In the latter case, a direct encoding was used. Neutral mutations were possible through the use of junk parts of the genome rather than through neutral mappings (see also Wu and Lindsay (Wu and Lindsay 1996) on the non-coding segments of DNA known as introns and Burke et al. (Burke, De Jong, Grefenstette, Ramsey, and Wu 1998) on the exploitation of variable length genomes for utilising junk DNA). These were sections of the genome that were functionless (having no effect on the fitness) in the current context of the rest of the genotype, but with different values elsewhere on the genotype may become useful. This may happen, for example, through gene duplication with a mutation on its promoter site deactivating that gene (i.e. a gene switch), allowing neutral mutations to take place on the duplicate, which may later become activated (Ohno 1970).

In their work on evolvable hardware, Harvey and Thompson (Harvey and Thompson 1996) found that the junk parts of the genotype with the highest potential for future usefulness coded for the periphery of the functional part of the FPGA, as they have the possibility of connecting newly evolved functional circuits to the existing active one to create a circuit of higher fitness. According to the authors, the larger this periphery the more potential there may be. In their experiments only neighbouring cells in the 2D layout were allowed, but the chip (Xilinx 6200 series) has the potential for "Magic wires" and "Fastlanes" which connect non-neighbouring cells, and in effect give a higher dimension topology for the chip than the simple 2-D layout (Harvey and Thompson 1996).

Another place where neutral mappings occur is in the mapping from the triplet sequences of bases, known as codons, in the DNA to amino acids. There are 64 codons, three of which are used to demarcate coding regions, while the other 61 map to the 20 naturally occurring amino acids. This gives a many to one mapping where most of the amino acids can be specified by many different codons. This occurs through a mechanism expounded in the "wobble hypothesis" (Crick 1966), whereby a mutation at the third position in a codon often codes for the same amino acid; hence a neutral mutation (in genetics this is usually referred to as a silent mutation). The genetic code is shown in Table 3.1, in Section 3.2.3.

Kargupta (Kargupta 2000) demonstrated that there exists a class of genetic code-like transformations that are able to convert a function of exponential description in Fourier basis to one with only a polynomial number of terms that are exponentially more significant than the rest when fitter proteins are given more copies through redundant and equivalent representation. Kargupta also found that natural selection alone may not be sufficient for efficient genetic search, redundant and equivalent representations may be required for increasing the proportion of fitter genotypes in the population.

### 3.1.4   Modularity of Development for Incremental Evolution

One important feature of developmental processes that has not been touched upon in evolutionary computation is its property of modularity. It is this property that allows the incremental evolution of form that is seen in nature.

It has been pointed out (Gilbert 1997) that only about three dozen animal body plans are currently being used on this planet. These, and the many long ago extinct phyla [1], were all formulated over 500 million years ago, with no new phyla emerging since. Kauffman (Kauffman 1993) proposes a mathematical model that predicts that any evolving system displays this pattern of divergence followed by the locking in on a particular subset of the original diversity. In a rugged fitness landscape, with all organisms starting with the same average fitness value (i.e. each starting midway up a peak), to begin with, if organisms take large jumps, they have 50% chance of becoming a fitter organism (finding a higher peak). Eventually, however, long jumps become risky as the chance of finding a higher peak (than the current one) decreases and the chance that these higher peaks are already occupied, increases. Instead, increasing fitness through small jumps (on the same peak) will tend to make the organism somewhat fitter than the surrounding population. Thus a diversification around a few successful models is what tends to emerge, as the duration between successful long jumps to higher fitness peaks doubles with each successive higher peak.

Although there are so few body plans, there are however, several million different species, each with its own pattern of development. Thus, most evolution has occurred within the framework of an existing body plan. The reason that the complex and finely tuned development of an embryo (its developmental pathway) can be altered (through evolution) at any stage of its development is because the development occurs through discrete and interacting modules (Gilbert 1997). Organisms are constructed of units that are coherent themselves and yet part of a larger unit. Thus cells are parts of tissues, which are

---

[1] In biological taxonomy, a phylum is the next division down from kingdom. For example, humans belong to the kingdom *Animalia* with phylum of *Chordata* and class *Mammalia*.

in turn parts of organs, and so forth. Modular units allow different parts of the body to change without interfering with other functions.

The principle of modularity allows four processes to alter development, these being dissociation, duplication and divergence, and co-option. Together, these allow incremental evolution to take place.

**Dissociation** When one part of the embryo changes, through mutation or environmental perturbation, without affecting other parts, this is known as dissociation. Such modularity of development allows changes both spatially and temporally to occur. Temporally, changes can occur as a shift in the relative timing of the developmental processes of modules (heterochrony), or as changes in rates of growth between modules (allometry). The latter can be very important in forming variant body plans (Gilbert 1997) such as seen in the differences in skull development between land-based mammals and whales, where in the latter the nose begins in the same place as a mammal's at the embryo stage, but due to the enormous growth of the upper jaw, is pushed to the top of the skull to become the whale's blowhole. Allometry can also generate evolutionary novelty with small incremental changes that eventually cross some developmental threshold. Eventually, a change in quantity becomes a change in quality. An example of this is the evolution of external fur-lined "neck" pouches of kangaroo rats and pocket gophers, which differ from internal pouches by having a fur lining, and no internal connection to the mouth. This allows seeds to be stored without risk of desiccation.

**Duplication and divergence** These are able to occur due to the modularity of development. Through duplication, redundant structures may be formed, and with divergence these structures can take new roles. One of the copies is able to maintain the original role while the others are free to mutate and diverge functionally. This may happen at the genetic level, through gene duplication, or at the developmental level, where copies of one structure diverge into a different structure during development (for example hair and teeth).

**Co-option** This occurs when pre-existing units are recruited for new functions. No one structure is destined for any particular purpose. This occurs on many levels: proteins may have different uses in different situations; genes may be used for segmentation, specification and even for providing an anterior-posterior axis; morphologically also, for example forearm structures may be modified to become wings.

One evolutionary consequence of the modular nature of development is that changes in one part of the developing organism (embryo) may induce change in another. This is known as correlated progression. An example of this is skeletal cartilage, which informs the placement of muscles, which in turn induce the placement of nerve axons. If one of these changes, it will induce the others to change with it (Thomson 1988).

## 3.2     Biological Background of Morphogenesis

In this section, the biological background of morphogenesis is introduced. First, the structure of a single cell is covered in Section 3.2.1, after which the processes involved in multicellular organism development are elaborated, in Section 3.2.2, of which morphogenesis itself is but one. The whole developmental process is driven by gene expression, and so this is then covered in some detail in Section 3.2.3. While gene expression drives development in each cell individually, inter-cellular signaling is responsible for coordinating these cells as a group, so that multi-cellular development is able to take place. This mechanism is covered last, in Section 3.2.4.

### 3.2.1     The Cell

Biological cells can broadly be classified into two different types, being prokaryotes and eukaryotes, determined by whether or not they have distinct subcellular compartments, most noticeably a nucleus in which the chromosomes containing the genetic information are located. Prokaryotes have no nucleus, and have a single chromosome. Bacteria, which are single celled organisms, are of this category. Eukaryotes have a nucleus which contains the chromosomes and genetic apparatus necessary to decode these. All higher level multi-celled organisms are of this type.

Figure 3.1 shows the general structure of a eukaryotic cell. The cell is enclosed by a plasma membrane, and in some cases (for example plants) a rigid cell wall, which allows the passing of small molecules. The nucleus is bounded by a double membrane called the nuclear envelope, which separates the contents of the nucleus from the cytoplasm and assists in regulating the flow of molecules into and out of the nucleus through nuclear pores. The region surrounding the nucleus and enclosed by the plasma membrane is known as the cytoplasm.

The cytoplasm contains all of the cell's organelles, the small sub-cellular structures, such as mitochondria, that perform dedicated functions. These are held in place, within the water-based cytoplasm, by a network of fibres throughout the cell's cytoplasm, known as the cytoskeleton. The cytoskeleton also gives the cell structural support, and helps the cell to maintain its shape.

Mitochondria are sites of cell respiration and are the main power generators in animal cells, converting oxygen and glucose to ATP (adenosine triphosphate), the cellular energy carrier, which is used for synthesising proteins from amino acids, the replication of DNA, growth and repair, and in muscle cells

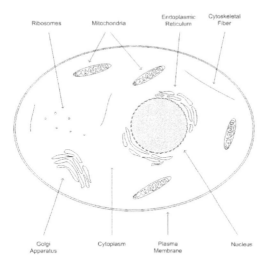

Ribosomes    Mitochondria    Endoplasmic    Cytoskeletal
                             Reticulum      Fiber

Golgi         Cytoplasm      Plasma         Nucleus
Apparatus                    Membrane

Figure 3.1: Structure of a Eukaryote Cell

for contraction.

Proteins are generated in the cytoplasm from messenger RNA (mRNA), originating in the nucleus. This is done on clusters of ribosomes (polyribosomes), or on the membranes of the endoplasmic reticulum. From here, proteins are transported to the Golgi apparatus which controls the storage and transportation of various proteins to where they are needed, which in some cases may be outside the cell.

## 3.2.2    Developmental Processes

Biologically speaking, development is the process by which a multicellular organism is formed from an initial single cell. The development process is progressive and gradual. From a single cell, a simple embryo comprised of a few cell types organised in a crude pattern is then formed, which is gradually refined to generate a complex organism comprised of many cell types organised in a detailed manner. This model of the development process is known as epigenesis, and is comprised of five major overlapping processes: *growth, cell division, differentiation, pattern formation,* and *morphogenesis* (Twyman 2001).

All of these processes are largely driven by and controlled by the expression of genes on the chromosome according the presence of gene-regulating proteins in the cell, and environmental signals. This mechanism, known as gene expression, is the process by which proteins signal cellular state to activate genes, which in turn are able to effect changes in cell state by the production of further proteins which may be used as signals or cellular building blocks. This provides a view of the cell as a set of parallel processing elements (genes) controlled by the interactions between their programs, encoded in the chromosome, and their environment, as represented by proteins detectable by the cell.

### 3.2.2.1   Growth and Cell Division

Growth and cell division may happen together or independently. In early animal development there is an increase in cell number without growth, so that the egg is divided into a series of progressively smaller cells. Later in development, cell division and growth occur together, although growth may occur without cell division through changes in cell size and depositing of materials such as bone into the extracellular matrix.

The extracellular matrix is a network of macromolecules secreted by cells into their local environment. Interactions between cells and the matrix can maintain epithelial sheets, provide a substrate for migration, and induce differentiation. The matrix is also the predominant component of some tissues, such as bone and cartilage.

Cell division influences organism development in two ways: the rate of cell division contributes to differential growth in different parts of the embryo; and the position and plane of cell division affects the size and orientation of the child cells.

### 3.2.2.2   Pattern Formation

Pattern formation is the process of organising and positioning cells in the embryo to generate an initially rough body plan which is then further refined to generate the more detailed structures of individual organs. This process firstly requires that differentiated cell types arise in appropriate places, which occurs through the regional specification, whereby cells specialise according to their location on various axes, which, in turn, is usually determined through chemical gradients, known as morphogens, that produce graded responses from cells according to concentration. Secondly, similar cell types need to form regionally-appropriate structures, such as bones.

The first pattern formed in the embryo is the principal body axis, which may be pre-determined by the distribution of maternal gene products in the egg, or require a physical cue from the environment. The specification of cells positions along each axis may involve morphogen gradients or developmental compartments.

Several pattern generating mechanisms that occur later, in organ development, utilise pre-existing patterns of gene expression. Examples of this include asymmetric cell division, lateral inhibition for generating regular spacing patterns and specific localised interactions between cells. Reaction-diffusion mechanisms, chemical systems comprised of multiple interacting diffusible components that are able to spontaneously generate patterns, are also suspected to play a role in the formation of patterns under some conditions (Twyman 2001).

### 3.2.2.3    Differentiation

Generally speaking, cells all contain the same genetic information, however, their specialised structures and functionality differ according to the proteins present within the cell, and this is determined by which genes are expressed. Differentiation is the process by which different patterns of gene expression are activated and maintained to generate cells of different types. Which genes, and hence proteins, are expressed differs between cells according to what cytoplasmic determinants are inherited at cell division, and what extracellular signals are received. Cytoplasmic determinants are molecules, such as transcription factors, that bind to the regulatory regions of genes and help to determine a cell's developmental fate (i.e. pattern of gene expression that causes the cell to differentiate in a particular manner).

Induction, whereby a signal received from another cell is able to affect the receiving cell's developmental fate, is used to control differentiation and pattern formation in development. An inductive signal may be used to instruct a cell to choose one fate over others, or be required to allow a cell already committed to a particular developmental pathway to continue towards differentiation. Inductive signals may occur over various ranges and may produce a single standard response in the responding cell, or a graded response dependent on signal concentration, in which case it is called a morphogen. Induction is covered in more depth in Section 3.2.4

### 3.2.2.4    Morphogenesis

Morphogenesis is the process by which complex structures, such as tissues and organs, are generated through the utilisation of cell behaviours. Cells are able to produce many individual and collective behaviours, such as changes of shape and size, cell fusion and cell death, adherence and dispersion, movements relative to each other, differential rates of cell proliferation, and cell-extracellular matrix interactions (Twyman 2001). Many of these behaviours have obvious developmental affects, while others utilise the properties of the biological medium. Changes in cell shape and size drive many folding and buckling movements, and can occur through the uptake of water or lipids into the cell, or due to reorganisation of the cytoskeleton. Cell fusion is used in developing muscle tissue, while cell death is responsible for creating cavities and gaps, such as required between digits in vertebrate limb development. Cell death also occurs in the developing nervous system through competition between neurons. Cell adhesion is responsible for tissue reorganisation through changes in the pattern of cell adhesion molecules. The expression of different classes of cell adhesion molecules helps to keep like cells together in tissues, and maintain boundaries between different tissues.

Morphogenesis is necessary for the embryo to progress beyond a simple ball of cells, and as such dynamic processes such as gastrulation, embryonic folding and organogenesis would not be possible without it. Early in the developmental process, morphogenesis is instrumental in driving the developmental program, for example by bringing cells together so that they may influence each other. Later in development, morphogenesis can be seen as being largely a response to the developmental program, for example causing an appropriate response from a differentiated cell according to its position.

### 3.2.3   Gene Expression

According to the central dogma of molecular biology, DNA (deoxyribonucleic acid) is *transcribed* into messenger RNA (ribonucleic acid), which is in turn *translated* into a sequence of amino acids, according to the *genetic code*, which are then constructed into proteins. These proteins are responsible for the generation of phenotypic traits as well as for regulating gene transcription (O'Neill and Ryan 2000).

A gene can be defined as being a functional unit of hereditary information. Structurally this corresponds to the *coding region* (of a gene), which is a segment of DNA that is involved in producing a protein (or possibly just RNA). A gene also includes *regulatory regions* preceding (*upstream*) and following (*downstream*) the coding region, and usually also has a *promoter* site to which *polymerase* binds for initiating transcription. In eukaryotes, there may also be intervening sequences (introns) between individual coding segments (exons), while in prokaryotes, there may be more than one coding region associated with a single promoter.

DNA is comprised of 2 complimentary chains of nucleotide bases. These nucleotide bases form a four symbol alphabet comprised of: Adenine (A), Thymine (T), Guanine (G), and Cytosine (C); of which G and C, and A and T form complements, such that each base on one strand of DNA will be bound to its complement on the other. The DNA in the coding region of a gene is transcribed into mRNA (messenger RNA), which acts as a photocopy of the gene by having a sequence identical to one strand of the DNA, but with Thymine (T) replaced with Uracil (U). This mRNA sequence transports the information stored in the DNA to ribosome in the cytoplasm, where it is used as a template for protein synthesis.

Each amino acid is specified by a particular combination of 3 nucleotides,

Table 3.1: The Genetic Code

| 1st base | | 2nd Base | | | | | | | | 3rd base |
|---|---|---|---|---|---|---|---|---|---|---|
| | | U | | C | | A | | G | | |
| U | Phe | UUU | Ser | UCU | Tyr | UAU | Cys | UGU | | U |
| | Phe | UUC | Ser | UCC | Tyr | UAC | Cys | UGC | | C |
| | Leu | UUA | Ser | UCA | Stop | UAA | Stop | UGA | | A |
| | Leu | UUG | Ser | UCG | Stop | UAG | Trp | UGG | | G |
| C | Leu | CUU | Pro | CCU | His | CAU | Arg | CGU | | U |
| | Leu | CUC | Pro | CCC | His | CAC | Arg | CGC | | C |
| | Leu | CUA | Pro | CCA | Gln | CAA | Arg | CGA | | A |
| | Leu | CUG | Pro | CCG | Gln | CAG | Arg | CGG | | G |
| A | Ile | AUU | Thr | ACU | Asn | AAU | Ser | AGU | | U |
| | Ile | AUC | Thr | ACC | Asn | AAC | Ser | AGC | | C |
| | Ile | AUA | Thr | ACA | Lys | AAA | Arg | AGA | | A |
| | Met | AUG | Thr | ACG | Lys | AAG | Arg | AGG | | G |
| G | Val | GUU | Ala | GCU | Asp | GAU | Gly | GGU | | U |
| | Val | GUC | Ala | GCC | Asp | GAC | Gly | GGC | | C |
| | Val | GUA | Ala | GCA | Glu | GAA | Gly | GGA | | A |
| | Val | GUG | Ala | GCG | Glu | GAG | Gly | GGG | | G |

called a codon. These amino acids make proteins, which make up cells - the building blocks of all complex living organisms. Although codons can encode for 64 ($4^3$) different amino acids, only 20 different types of amino acid are actually produced, providing redundancy of representation. Codons are allocated sequentially in the mRNA, with translation into amino acids beginning on a start codon (AUG) and ending when a stop codon (UGA, UAA, or UAG) is reached. The genetic code is given in Table 3.1 (adapted from (Winter, Hickey, and Fletcher 2002)).

The resulting chain of amino acids is used to construct the three dimensional structure of folded proteins. Although amino acid sequences essentially define proteins, the formation of the three dimensional structure of the protein involves a complex process known as protein folding. This process involves interaction between multiple amino acid sub-sequences, which results in the emergence of a folded protein structure.

There are two particular types of protein that have important roles in development, these being transcription factors (TFs) and components of signaling pathways. Transcription factors control gene expression by binding at sites on the regulatory regions of a gene, hence playing a major role in the coordination of developmental processes; whereas signaling pathways are necessary for cells to be able to perceive external signals.

Control of gene expression can take place at any of the intermediate stages of transcription, RNA processing, mRNA transport, mRNA degradation, and protein activity. Transcription of DNA, whereby the coding region of a gene is transcribed to an RNA molecule prior to translation into a protein, is the first level of regulation of gene expression and hence the developmental process, and is generally the most important level of control in gene expression (Reil 2000; Twyman 2001). Transcriptional regulation involves regulating the rate at which the RNA polymerase enzyme transcribes the gene into mRNA.

Prokaryote genes are transcribed in the cell's cytoplasm by RNA polymerase to generate mRNA. This begins at the promoter site and ends at a terminator site, downstream from the coding region. The terminator site contains self-complimentary sequences which form a stem-loop or hairpin structure that stops the polymerase from continuing transcription. Within the promoter are located short conserved sequences, where polymerase binds. Differences in these sequences give rise to different transcription initiation efficiency, which contributes to gene regulation. Prokaryote mRNA transcripts may contain more than one coding region (encoding a protein), enclosed within start and stop codons. The encoded regions of mRNA are translated into proteins by ribosome, using transfer RNA (tRNA) to deliver amino acids, as specified by the genetic code, for constructing the protein.

Eukaryotes differ in that transcription takes place in the cell's nucleus, while translation occurs in the cytoplasm. Also, in eukaryotes, mRNA generally only encodes a single protein, but the protein encoding region often has intervening non-coding regions, known as introns, which are removed in the pre-mRNA, while the remaining coding regions (known as exons) are spliced together to form a single mRNA.

Eukaryotes are more complex than prokaryotes, and have 3 different types of RNA polymerase. RNA Polymerase II is responsible for transcription of mRNA. Eukaryotes also generally have a promoter, known as the TATA box, which is comprised of the DNA sequence $TATA(A|T)A(A|T)$, to which basal transcription factors bind to create a platform which is used to position RNA polymerase II for transcription. Eukaryote promoters may have their efficiency increased through upstream regulatory elements (UREs) located within 100-200 base pairs of the promoter (Twyman 2002), and may also be influenced by regulatory sequences up to thousands of bases distant, known as enhancers, when they act to increase the transcription rate, or silencers, when they act to decrease the transcription rate.

Regulation of gene transcription occurs by the binding of transcription factors at specific sites. In prokaryotes, gene transcription may be inhibited by the binding of transcription factors overlapping the promoter sequence or between the promoter and the coding region, thus blocking transcription. There is also considerable variation in sequence between different promoters, caus-

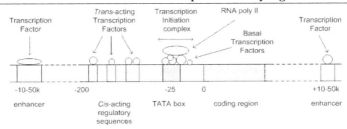

Figure 3.2: Structure of a Eukaryote Gene

ing variations in transcriptional efficiency by up to 1000-fold. In some cases transcription requires binding of a transcription factor to an enhancer region upstream of or overlapping with the promoter for efficient transcription to occur.

Transcription factors may themselves be disabled (from binding at regulatory regions) by proteins binding to them, as in the LAC operon (see following section), or may require the assistance of another protein to enable binding, as in the TRP operon.

In Eukaryotes, RNA polymerase II requires the presence of additional initiation proteins before they are able to bind to promoters for initiating transcription. Eukaryotic genes have minimal rates of transcription that are greatly improved by the binding of transcription factors to UREs and enhancers. Repressor transcription factors may also block transcription in the same manner as prokaryotes, by binding to transcription factors responsible for activating the gene, causing them to be disabled, or by blocking transcription. Several other regulatory mechanisms are also available. Figure 3.2 illustrates the general structure of a eukaryotic gene (adapted from (Winter, Hickey, and Fletcher 2002)). Here, the *cis*-acting regulatory sequences, are the UREs mentioned above, while the *trans*-acting factors are specific transcription factors that bind to these.

Prokyarotes provide a simpler conceptual model of the basic mechanisms involved in gene expression. For this reason, the operon model of gene expression in prokaryotes is presented below.

### 3.2.3.1    Operon Model of Gene Expression

In 1961 Jacob and Monad proposed the operon model for the co-ordinate regulation of transcription of genes involved in specific metabolic pathways in bacteria. An operon is a collection of linked genes located contiguously on a stretch of DNA, forming a single strand of messenger RNA, and is under the control of one promoter, to form a unit of gene expression. It is typically comprised of structural and regulator genes, and control elements such as enhancer (positive regulation) and/or operator (negative regulation) DNA sequences that regulate transcription of the structural genes. Regulator genes produce products that bind to the control elements, while the structural genes are genes code for structural proteins or enzymes that don't contribute to the operon's regulation.

The LAC operon, found in the E. coli bacterium, provides a good illustration of the operon model. E. coli is able to metabolise either glucose or lactose. When there is no lactose, the *lac* operon is inactive, being disabled by the *lac* repressor, which is encoded in a gene upstream from the operon. The *lac* repressor binds to the operator, located downstream of the promoter. When the repressor is bound to the operator, RNA polymerase is unable to proceed downstream with its task of gene transcription.

However when the bacterium comes into contact with lactose, the minimal amounts of the *lac* enzyme present in the cell allow some lactose to be metabolised, and in the process allolactose is produced. Allolactose binds to the *lac* repressor, altering its conformation, so that it is no longer able to bind to the operator. This unblocks the RNA polymerase, which is then able to transcribe the structural genes, which are involved in the metabolism of lactose.

When both lactose and glucose are present, glucose is the preferred energy source, and so, while absence of the *lac* repressor is essential for transcription to occur, it is not sufficient for effective transcription of the *lac* operon. The activity of RNA polymerase also depends on the presence of another DNA-binding protein called catabolite activator protein (CAP), which binds to the enhancer site upstream of the promoter. However, CAP can bind to DNA only when cyclic adenoside monophosphate (cAMP) is bound to CAP, and cAMP production is inhibited by the presence of glucose in the cell. Thus, when cAMP levels in the cell are low, indicating that glucose is present, CAP fails to bind DNA and transcription of lactose metabolising enzymes is inhibited, even though the repressor is not active.

The general structure of the *lac* operon, which is under both negative (repressor) and positive (enhancer) control, is shown in Figure 3.3 (adapted from (Winter, Hickey, and Fletcher 2002)). Here, the *lac* repressor is bound by allolactose, disabling its ability to bind to the operator, while CAP is bound to by cAMP, enabling it to bind to the enhancer, thus encouraging transcription of the *lac* genes.

The operon mechanism is characteristic of prokaryotes, but differs in several respects from that usually found in eukaryotes. Genes in eukaryotes are not linked in operons; gene transcripts (mRNA) in eukaryotes contain the transcript of only a single gene; and transcription and translation are not physically linked in eukaryotes, as transcription occurs in the nucleus while translation occurs in the cytoplasm (Kimball 2002).

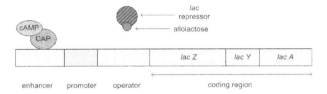

Figure 3.3: Structure of the *lac* Operon

## 3.2.4    Inter-Cellular Signaling

Inter-cellular signaling is the mechanism that enables cells to coordinate their behaviours, instantiated as patterns of gene expression, through a variety of externally recognisable signals. Briefly, cellular signaling involves the following sequence of activities. When a, typically molecular, signal arrives at the cell it binds to a membrane-spanning receptor on the cell surface, which in turn activates signaling pathways within the cell, resulting in a change in the cell's cytoskeletal organization, ion permeability, enzyme activity or gene expression.

The core processes involved in inter-cellular signaling are the transfer and interpretation of an extra-cellular signal to an inter-cellular signal, known as *signal transduction*, and the modification of the receiving cell's behaviour through *induction*. These are covered in turn below.

### Signal Transduction

To be able to perceive an external signal, the receiving cell needs the signal to be transferred to its interior; signal transduction is the mechanism by which this is achieved. In most cases, the external signal acts as a ligand (a small molecule that binds to a larger macromolecule) for a membrane-spanning receptor on the cell's surface.

When a ligand binds to the extracellular ligand-binding domain of its receptor, it induces a change in the conformation of the receptor's cytoplasmic domain, which in turn stimulates a cascade of kinase (any of the various enzymes that catalyse the transfer of a phosphate group from a donor to an acceptor) activity resulting in the activation of dormant signaling proteins (transcription factors) within the cell (Twyman 2001). In this manner, an extracellular signal is converted into a different type of intracellular signal without the having to enter the cell.

Cells respond to diverse range of external signals, including physical stimuli, such as light and heat, molecules in the environment, and contact with other cells. Cells have a smaller repertoire of responses than there are stimuli, resulting in the signal receptors channeling incoming signals into common signal transduction pathways.

In general, signaling pathways consist of a series of proteins, each of which acts on the next to alter its conformation, and so alter its activity. This is achieved by sequentially adding (by kinases) and then removing (by phosphatases) the phosphate groups of the target molecules, to create a signal path (Karp 1998).

The cell's response to a signal depends on which signaling components, transcription factors and other proteins are available in the cell. This allows the same types of signal and receptor to be used over and over again in different developmental systems. The result of this is that many signal transduction pathways are highly conserved in different biological systems and in different organisms.

Signaling pathways don't operate in isolation; they are interconnected at many levels, and are regulated by feedback and cross-talk from other pathways, so that cells respond to the sum of signals arriving at their surfaces, without becoming overloaded with information. Hence, signaling pathways are controlled by a complex system of feedback and cross regulation that reflects the balance of opposing forces in the cell. The arrival of a signal at the cell surface causes a disruption to the equilibrium, allowing a brief pulse of information transfer, before the equilibrium is restored. In this manner, cells are able to compensate for changes in their environment while maintaining homeostasis (Twyman 2001).

**Induction**

For cells to be able to affect the patterns of gene expression in other cells, and hence their developmental fate, requires that a molecular signal secreted by one cell is perceived and acted upon by the responding cell. This form of inter-cellular communication is known as induction, and is a common strategy for controlling cellular differentiation and pattern formation.

In development, definitive changes accompany inductive interactions, reflecting the strong and unambiguous nature of inductive signals. Thus allowing changing patterns of gene expression to establish a new equilibrium in the cell (Twyman 2001).

Induction involves cell to cell signaling, that may occur over various ranges, and requires that an inducing cell is able to generate the signal and along with the ability of the destination cell to receive and react to the signal. The ability of the receiving cell to respond to an inductive signal in an appropriate manner is known as competence. The loss of competence is one of the mechanisms by which cells become irreversibly committed to a given developmental pathway.

There are two types of induction, instructive and permissive, based upon the choices available to the responding cell. Instructive induction is where the receiving cell has a choice of developmental fates, and the inductive signal instructs the cell to choose one fate in preference to others, while permissive induction occurs when the responding cell is already committed to a certain developmental pathway, but needs the inductive signal to be able to continue towards cellular differentiation.

Cells may have a single stereotypical response to an inductive signal, or a graded response dependent on signal concentration. In the latter case, the signal is known as a morphogen, and may be used to produce different cell responses based on the distance from the morphogen's source, allowing positional information to be communicated.

Cell populations may also respond differently to individual cells to inductive signals. For example, in the community effect a developmental fate that can't be taken by isolated cells can, however, be undergone by a population through the increased concentration of the inducing signal. Lateral inhibition is another example of this, whereby a collection of originally identical cells are able to differentiate into regularly spaced patterns of different cells due to the inhibitory signals sent between adjacent cells.

Through these inductive mechanisms, cells are able to differentiate into various cell types with specialised structures and functionality, according to their location and relationship with other cells within a broader body plan.

## 3.3    Morphogenesis in EHW and EC

There are a few general approaches available for implementing models of morphogenesis in EHW. One popular method is known as generative encodings, and is typically implemented with Lindenmeyer systems (commonly referred to as L-systems). There have also been a few attempts at modelling cellular biological processes, driven by gene expression (Gordon and Bentley 2002b; Roggen, Floreano, and Mattiussi 2003; van Remortel, Manderick, and Lenaerts 2004).

### Generative Encoding

Morphogenesis can be achieved in evolutionary computation through the use of a generative encoding scheme. This involves the use of a grammatical encoding in which the encoding specifies how to construct the phenotype. It is similar to a computer program, in that it allows for the definition of sub-procedures, which allows the system to scale up to higher complexity than can be achieved with a direct encoding (Hornby, Lipson, and Pollack 2001).

The idea of incorporating the development process to map a genotype into a phenotype was proposed by (Kitano 1990) for evolving a neural network structure, using an L-system as a graph rewriting scheme. Instead of acquiring a connectivity matrix directly, the generative encoding method obtains rewriting rules from which the connectivity matrix can be generated by successive application of these.

It has since been shown that a generative encoding scheme can achieve greater scalability through self-similar and hierarchical structure, and in addition, by reusing parts of the genotype in the creation of the phenotype, a generative encoding is a more compact encoding of a solution (Hornby and Pollack 2001a).

**Cell-based models driven by Gene Expression**

Another way of implementing morphogenesis in evolutionary computation is to emulate the cellular mechanisms involved in biological development. This typically involves using a model of gene expression to drive the developmental process in a set of cells, along with some form of inter-cellular communication to coordinate the morphogenesis process across multiple cells.

Gene expression models offer several advantages over standard genetic algorithms:

- Separation of search and solution spaces (genotype-phenotype mapping).

- Compression of representation by encoding growth instructions.

- Variable length genotypes allowing a variable number of genes (rules) and gene expression constraints (pre-conditions).

- Degenerate (many-to-one) encodings providing efficient search through neutral mutations.

- Maintenance of genetic diversity and preservation of functionality while allowing continuation of search through neutral mappings and junk DNA.

- Alternative implementations of functions, through an operon model for example.

- Positional independence, which may allow a genetic algorithm to adapt the structure of the solutions being generated, for example by finding advantageous positions of co-dependent genes for preventing deleterious recombinations (Burke, De Jong, Grefenstette, Ramsey, and Wu 1998).

- A generalised encoding that can represent a wide variety of structures without requiring specially constructed genetic operators (the functional diversity of proteins) (O'Neill and Ryan 2000).

**Outline**

The rest of this section provides details of these two main approaches to morphogenesis in EHW and EC. Firstly, L-systems are covered, and then their use in EHW is reviewed. Following this, cell-based models in EHW are reviewed, and lastly computational models of gene expression that have been used in EHW or EC in general are covered.

### 3.3.1   L-Systems

Lindenmeyer systems (L-systems) were introduced by (Lindenmeyer 1968) as a mathematical formalism for describing the development of branching structure found in many plants. For this reason the initial L-system notation simply represented graph-theoretical trees. Subsequently, geometric interpretations were proposed to allow L-systems to be used for presenting graphical models of plants and fractals in particular (see (Prusinkiewicz and Hanan 1989) for a complete presentation on this area).

L-systems provide a necessary and convenient distinction between genotype and phenotype, and a well defined process (morphogenesis) to generate the latter from the former (Ochoa 1998).

A beauty of L-systems as a generative encoding is that it is a general generative encoding system; by changing the language of terminals, different structures can be generated, such as plants (Prusinkiewicz and Lindenmeyer 1990), furniture (Hornby and Pollack 2001a), artificial neural networks (Kitano 1990), evolvable hardware (Kitano 1996b), and locomoting creatures (Hornby, Lipson, and Pollack 2001; Hornby and Pollack 2001b).

Other advantages of L-systems are that they are syntactically closed and well conditioned under genetic operators. Informally, the latter requires that "small" mutational changes should (usually) cause "small" phenotypic changes, and that crossover usually produces offspring whose phenotypes are in some sense a "mixture" of the parents' phenotypes, with occasional jumps and discontinuities (Ochoa 1998).

#### Categories of L-Systems

The main categories of L-systems are:

- Simple. These only allow the parallel rewrite of symbols.

- Bracketed. These also allow branching tree-like structures to be formed by using brackets as push and pop operators.

- Map (or Graph). These allow structures of segments and enclosed space (as in cellular structures which are enclosed by walls) to be represented through the rewriting of parallel graphs with cycles.

- Simple, Bracketed, and Map L-systems are also known as deterministic context-free L-systems, denoted as D0L-systems.

- Context-sensitive. These also take into account the preceding and succeeding symbols.

- Parametric. These are similar to context-free, but also take parameters; denoted as P0L-systems.

- Probabilistic. Rather than having conditions these have multiple successors, each with a probability of being activated.

### Context-free L-System Example

As an example of a simple (context-free) L-system, consider the following (taken with alterations from (Green 1993)):

Given the rules: $A \rightarrow B; B \rightarrow AB$

and an initial state of $A$, this L-system produces the following sequence of strings:

Stage 0 : $A$
Stage 1 : $B$
Stage 2 : $AB$
Stage 3 : $BAB$
Stage 4 : $ABBAB$
Stage 5 : $BABABBAB$
Stage 6 : $ABBABBABABBAB$
Stage 7 : $BABABBABABBABBABABBAB$

Counting the length of each string, gives the Fibonacci sequence ( 1, 1, 2, 3, 5, 8, 13, 21, ... )

### Context-sensitive L-Systems

0L-systems are context-free, meaning that they are applicable regardless of the context in which the predecessor appears. This is sufficient for information transfer between generations, but is unable to model information exchange (interactions) between neighbours, for which a context-sensitive extension to L-systems is required (Prusinkiewicz and Hanan 1989).

These extensions may have context to the left, right, or both. In the first and second case, these are known as 1L-systems, and the last case is a 2L-system. These belong to a wider class of IL-systems called (M,N)L-systems, in which the rewriting of a letter depends on M of its left and N of its right neighbours, where M and N are fixed integers. These resemble context sensitive Chomsky grammars, but as L-system rewriting is parallel in nature, every symbol is rewritten in each derivation step. This is especially important whenever there is an overlap of context strings. While this class of L-systems is deterministic, the lack of parameters results in the inability to take advantage of parametric terminals.

## 1L-System Example

An example (taken from (Prusinkiewicz and Hanan 1989)) for which a 1L-system can be used is for modelling hormone diffusion along a filament. If '$a$' denotes a cell with hormone concentration below a threshold level, and '$b$' a cell with its concentration exceeding this level, then diffusion process can be described with the rule: $b < a \rightarrow b$. Starting with the string "$baaaaaa$", the following is produced:

*baaaaaa*
*bbaaaaa*
*bbbaaaa*
*bbbbaaa*
*bbbbbaa*
*bbbbbba*
*bbbbbbb*

## Parametric and Probabilistic L-Systems

With Parametric L-systems production rules also have parameters, with which algebraic expressions can be applied when passing parameter values to successors. Also, parameter values can be used to determine which production rule to apply.

Probabilistic L-systems' production rules do not have a condition part, instead they have multiple successors, each with a probability that it will be used to replace the predecessor. While good for generating a variety of similar structures, this system is non-deterministic, making it unsuitable for developing structures that need to be recreated the same each time.

## L-Systems Applied to Evolving Complex Structures

Much work has been done on using L-systems in evolving plant-like structures. Some examples include (Jacob 1994; Jacob 1996; Ochoa 1998).

Kitano (Kitano 1990) introduced the idea of evolving neural networks using an L-system on matrices to generate the connectivity matrix for the network. Other examples of evolving neural networks using L-systems include (Lee, Lee, and Sim 2000) and (Channon and Damper 1998).

Work has also been done on co-evolving robot morphology and controllers. (Hornby, Lipson, and Pollack 2001) coevolved 2D robots with motorised joints each controlled by an oscillator, and successfully transferred these from simulation to real robots.

In (Hornby and Pollack 2001b) the authors extended their previous work further, coevolving 3D simulated creatures with neural network controllers. The authors' system creates networks in a method similar to that of cellular encoding (Gruau 1994), with operators acting on links instead of the nodes.

Whereas the generative encoding of (Sims 1994) allows for repetition of segments, it did not produce hierarchies of regularity. Hornby and Pollack's generative encoding system (Hornby and Pollack 2001b) is a more powerful language with loops, sub-procedure-like elements, parameters and conditionals; and achieves an order of magnitude more parts than the previous work of (Sims 1994).

One of the most demonstrative works performed, showing the power of generative encoding in general, and L-systems in particular, for evolving complex systems is found in (Hornby and Pollack 2001a). The authors evolve L-systems to generate table designs, using parametric context-free (P0L) systems, which they state is a more powerful class of L-systems than previously evolved. The authors reported that, even using different weightings and variations of the components of the fitness function, the generative encoding produced better tables than the non-generative encoding in all cases. On evolutionary runs with the non-generative encoding, over half the trials converged to poor structures and no run produced structures with complex regularities. In contrast, most trials using the generative encoding converged to good structures, and even those with low fitness had regularities.

The difference between the non-generative encoding and the generative P0L-system encoding are block replication and parametric production rules. Variants of the generative encoding, using only one or other of block replication or parametric production rules, did worse than the original generative encoding, but better than the non-generative encoding. Both variants produced tables with regularities, although the variant lacking block replication didn't necessarily have the sequential replications of structures as did the variants having it.

Block replication and production rules with parameters differentiated the work of (Hornby and Pollack 2001a) from previous work in evolving L-systems, although both features are similar to those of past work in evolutionary design and have analogues in computer languages.

Block replication is similar to for-next loops in programming languages, and is almost identical to the multiple re-writing of the recurrent symbol (using the life register) of cellular encoding (Gruau and Quatramaran 1996) and the recursive-limit parameter in graph encoding (Sims 1994).

Production rules are like subroutine calls in programming languages and are similar to the automatically defined sub networks (ADSN) of cellular encoding and automatically defined functions (ADFs) of genetic programming. With analogues to loops and parameterised subroutines, evolution of generative encodings becomes like the evolution of a computer program, as in genetic programming.

## 3.3.2    L-Systems in EHW

Early work at applying L-systems to EHW was based on work done with neural networks. Neural networks and logic circuits have several features in common. Both can be represented with graphs (any configuration can be represented using arcs and nodes), and in both cases functional local structures can be identified. Local structures are important in the design of complex circuits, being used in various combinations to achieve the desired functionality. This implies that the ability to move, copy and combine local structures is critical to evolving scalable circuits (Kitano 1996b).

### Early Work

As an approach to evolvable hardware Kitano (Kitano 1996a; Kitano 1996b) used a similar approach to his earlier work on neural networks (Kitano 1990) to evolve circuits for an FPGA.

Kitano's approach to EHW used a graph L-system as an approach to evolving circuits. Each chromosome contained a set of rules which were successively applied to rewriting the graph until the desired size was obtained. A rule in this approach is represented as:

```
LHS -> [ UL UR ]
       [ LL LR ]
```

This rule is applied when the left hand side (LHS) matches a symbol in the matrix. The symbol is rewritten as a 2x2 matrix with 4 symbols. For example a fragment of the chromosome, AOHCL, represents the following rule:

```
A -> [O H]
     [C L]
```

Although the alleles of the chromosome are represented here as letters, in the implementation Kitano used an integer value. At each allele position, an integer ranging from 0-19 was assigned; presumably meaning that up to twenty rules can be encoded. Thus, Kitano claims, the chromosome can encode for more than N rules, where N is the number of non-terminals, allowing redundancy, which better ensures that another rule can be used when a rule is destroyed by mutation or crossover. In other words, it seems that Kitano allows for multiple productions with the same left-hand side (as shown in the rule set given in (Kitano 1996b), Fig. 2), although presumably only one of these is activated.

The rewriting process starts from an initial node. For each symbol in the matrix, the matching rule is applied and the rewriting cycle continues until a matrix of the required size has been generated.

Kitano used this approach to evolve several combinatorial logic circuits on a logic simulator; however, these were not implemented on an actual FPGA, which would require an extra step of cell placement and routing. What was evolved were a set of logic gates (AND, OR, NOT, XOR, NAND, NOR) and the connections from the input receiving gates to the final gates.

Experimental results demonstrated that this approach consistently outperformed the direct encoding method (which evolves the circuit matrix directly) (Kitano 1996b).

### On a Xilinx 4000 Series FPGA

Early work on applying L-systems to generating circuits on an actual FPGA was done by Stauffer and Sipper (Stauffer and Sipper 1998). In this work, L-systems were used to model cellular development visualised as a cellular automata, which was then implemented on a Xilinx 4010 FPGA.

### On a Virtual Xilinx 6200 Series FPGA

More recently, (Haddow and Tufte 2001; Haddow, Tufte, and van Remortel 2001) have presented work on using morphogenetic process to evolve circuits on the Xilinx Virtex family of FPGAs. This approach was taken to reduce the complexity of the genome, thus allowing more complex circuits to be evolved. The mapping from genotype to phenotype was split into two stages, using a virtual EHW architecture as an intermediate stage between the genotype and the implementation on the Virtex device.

More recently Haddow, et al. (Haddow, Tufte, and van Remortel 2001) presented work using L-systems to generate circuits on a virtual Xilinx 6200-like architecture consisting of a matrix of 'sblocks' (see (Haddow and Tufte 2001) for a detailed description of the architecture).

In this work, binary production rules were used, with the left-hand side of the production being matched against the binary configuration bitstream of the virtual FPGA, either to match the state of a cell, or that of neighbouring (in the bitstream) cells. When a match is found by the left-hand side of a production rule, the right-hand side produces either a change of cell state or cell growth. Cell growth is limited, however, to cells with at least one unconfigured neighbouring cell. Cell growth results in placing a new sblock in the first free neighbouring cell location, in the order of N, S, E, W.

This scheme was used in an attempt to evolve routing blocks, cells whose only function is to transfer signals across cells, with only some limited success being achieved when crossover was removed to eliminate the effect of epistasis properties.

Epistasis refers to the suppression of a gene by the effect of an unrelated gene. Hence, crossover is able to disrupt the expression of a gene (a production rule in this case) due to the elimination of another gene whose product (in nature a protein, but here a change of cell state or cell growth) was required to trigger it.

The authors stated that their achieved solution was closer to hill climbing than evolution (Haddow, Tufte, and van Remortel 2001). In later work (Tufte and Haddow 2003) the authors reduced the search space by limiting evolved S-block functions to those considered useful to achieving their task, by adding neighbour information to the rules applied to an S-block to give local knowledge to the development process, and by introducing a cell-death feature. However, no results were presented on the usefulness of applying this to evolving a solution to an EHW problem, such as evolving routing blocks as in their earlier paper.

**Local String Alignment for Analog Networks**

A related approach to L-systems, using local string alignment, was used by Mattiussi and Floreano (Mattiussi and Floreano 2004) for generating analog networks. This was based on associating character strings extracted from the chromosome with component terminals and parameters, and generating the connections between these components using local string alignment. This approach was applied with some success to evolving a voltage reference circuit, implemented in a SPICE simulation. However, the authors concluded that there appeared to be no natural way of implementing the evolution of parametised component values and circuit topologies, as can be done with GP.

### 3.3.3    Cell-based Models for EHW

An alternative to using production rules for implementing development is to use a more biologically realistic cellular model, in which development is driven by gene expression. Here, genes code for proteins (the building blocks of biological organisms) and the expression of genes is regulated by the presence of proteins in the cell, some of which may be the result of protein signals from other cells. There have been some examples of this approach for evolving neural networks for robot control (Jakobi 1995; Eggenberger 1996; Roggen, Floreano, and Mattiussi 2003), EHW (Gordon and Bentley 2002b; Roggen, Floreano, and Mattiussi 2003; van Remortel, Manderick, and Lenaerts 2004; Liu, Miller, and Tyrrell 2005b), 2D structures (de Garis 1999; Bentley and Kumar 1999) and 3D organisms (Eggenberger 1997b).

In the application of cellular models to evolutionary methods, such as EHW, proteins are encoded in genes in the chromosome, which are activated through their regulatory regions, by proteins in the cell. This concept, as applied to

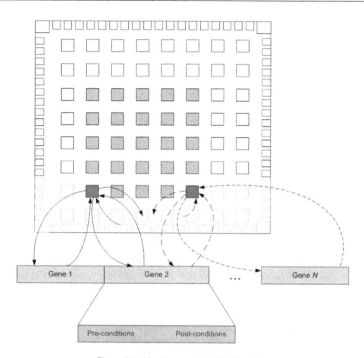

Figure 3.4: Morphogenetic EHW Encoding

EHW on an FPGA, is illustrated in Figure 3.4.

To initiate development, some cells may be seeded with particular proteins to activate gene expression. Cells continue to interact and develop for a number of developmental steps to give a resulting structure. This process of generating a phenotype (structure) from a genotype reflects the process of morphogenesis as seen in nature.

A review of EHW implementations of morphogenesis using cell-based models are presented next, while a review of gene expression models is covered later in Section 3.3.4.

### Xilinx Virtex Implementation

Gordon and Bentley (Gordon and Bentley 2002b) used a cellular model for evolving adders on a Xilinx Virtex. This was done by evolving cellular automata rules on a matrix of cells, each of which maps to a distinct CLB on the

72

FPGA. In this work, cells have five components, three of which map to components of the CLB, these being the inputs, outputs, and function generators (LUTs); and another two which exist only on the evolving cells, these being protein generators and detectors.

Developmental rules are binary strings of fixed length specifying the rule precondition and postcondition. The precondition specifies the required presence or absence (or don't care) of each of the five proteins for the postcondition to be activated. The postcondition is a key to a lookup table specifying the action to take: either generation of a protein or selection of cell LUT inputs, function and outputs.

Each cell contains two LUTs, F and G, as contained in a Virtex CLB slice, each with independent outputs. These cells map directly to a Virtex CLB slice, with the other slice, carry logic and flip flops disabled.

The postcondition doesn't directly allow selection of a LUT's 4 inputs, instead postconditions are able to increase the score of any particular input, and the selection is done through a competition between the 8 neighbouring cell outputs (each cell has two LUT's each having a single output). LUT's functions are controlled with postcondition keys (P-terms) for each of the 16 (1 bit) truth table entries per LUT; a score being kept of how many times each P-term has been activated, and a threshold being applied. If the score exceeds the threshold then the LUT entry is set to true, otherwise false. The threshold is dependent on the number of rules and proteins, being set to the expected P-term score if a set of random rules were activated. (Another approach was used with a Karnaugh map and move and set commands within that map. This had lower performance than the threshold approach however). Each LUT's single output is dealt with in similar manner using a threshold.

When this scheme was applied to evolving two bit adders with carry, it was found that it didn't perform as well as a non-generative encoding scheme, being unable to find a fully functional solution. The authors attributed this to the increased epistatic interactions (that is the suppression of a gene by the effect of an unrelated gene) that may occur with an additional mapping from genotype to phenotype, if a phenotypic feature relies on the presence of more than one gene product. Additionally the fixed rule lengths here mean that building blocks composed of more than one gene product require multiple rules, thus increasing their probability of disruption by crossover.

The biggest impediment to evolution, according to the authors, is that evolution is searching bias space. This means that evolution must find a representation that is evolvable before evolution is able to successfully search for a solution. However, the author's results with random search seem to indicate that the density of better solutions may be higher in developmental space than in a non-developmental encoding space.

In a later paper (Gordon 2003) it was found that some of the problems were due to the developmental model in which a cell's protein environment consists only of proteins generated by at least half their neighbours, and the lack of the concept of protein concentrations. When these changes were implemented, and protein diffusion added by giving locally produced proteins higher concentrations than those of neighbouring cell productions, although their system was unable to produce any perfect solutions in 20 runs of 5000 generations, they were finally able to produce 2 perfect solutions from 20 runs of a hill climbing algorithm over 500,000 evaluations, with point mutation of 5 per chromosome and new equal-to-best-fitness candidates replacing the existing fittest with a 50% probability (Gordon 2003).

More recently this model has been applied to investigating scalability of development for evolving n-bit (for n=1..7) adders with carry and even n-bit parity circuits (for n=2..12) on a simplified virtual FPGA architecture (Gordon and Bentley 2005). In both problem sets it was demonstrated that a developmental process enhances scalability. Furthermore, analysis of circuits revealed a high degree of design re-use by evolution, but when evolution was biased towards a standard decomposition of the problem, performance was lowered. This demonstrates that evolution is often able to find better design abstractions than those preconceived by a human designer.

**The POEtic System**

Another example of a cell-based approach is the work of Roggen, et al. (Roggen, Floreano, and Mattiussi 2003) implemented morphogenesis using mechanisms inspired by gene expression and cell differentiation to evolve spiking neural networks. This was implemented on a specialised Bio-inspired hardware system, known as the POEtic system, which is comprised of a 2D array of functional units that have the ability to evolve, self-repair and grow, and learn (Tyrell, Sanchez, Floreano, Tempesti, Mange, Moreno, Rosenberg, and Villa 2003).

Their morphogenetic process works in 2 phases: first a signaling phase in which adjacent cells exchange signals, resulting in a diffusion of signals across the cellular matrix, with signal intensity decreasing with the Manhattan distance from the source cell; and followed by an expression phase where the cell functionality is set by matching cell signal intensities to the closest (by hamming distance) functionality given in an expression table. Each cell contains a spiking neuron whose functionality is the connection type (there are 6 available) and characteristic (excitory or inhibitory). The genetic code used to evolve this contains two parts: the expression table containing 12 entries, one for each type of neuron, with each entry holding the 4 different diffusion signals and their signal intensities; and the 16 diffusion cell coordinates and their types.

This system was used to evolve neural networks for pattern recognition and mobile robot control. In both cases it outperformed a direct encoding, when comparing the maximum fitness reached and when comparing the number of evolutionary runs that were able to produce individuals with the maximum fitness.

A comparison of random walks from the fittest individuals showed that the fitness drops faster with the morphogenetic system, than with a direct encoding as it moved farther from points of maximum fitness, seeming to indicate that the fitness landscape is more rugged with the morphogenetic approach. Also, from measuring large populations of randomly generated individuals, the morphogenetic approach tended to give higher fitness values and standard deviations than the direct encoding approach. The authors surmised that the higher fitness variability might explain why the morphogenetic approach was able to generate individuals of the same fitness faster than a direct genetic encoding.

### Simple Gene Regulated Development

A morphogenetic model that has been proposed for future application to EHW is given in (van Remortel, Manderick, and Lenaerts 2004). A simple gene regulation model, where each gene has a single precondition and a single gene product and the chromosome is evaluated from left to right, was combined with a multicellular model, with the eventual aim of applying this to EHW. This model uses a lattice of cells, with inter-cellular communication between cardinal neighbours. Cells are able to divide asymmetrically, grow, move, age and die. Cells contain protein concentrations and inherited nuclear determinants and have structural properties (that can be used to define circuits), all of which can be used to direct gene expression. The authors did not present any work applying this model to EHW yet, however.

### Development Model for Digital Systems

Another cellular developmental model for EHW is detailed in (Liu, Miller, and Tyrrell 2005a) aimed at generating robust combinatorial digital circuits. In this model, each cell has a 2-bit state value (cell type) and contains a 4-bit chemical state. The cell's state at each growth step is determined by its own state and chemicals and those of its four adjacent cells. Each cell takes inputs from its west and north neighbouring cells and provides its calculated output to its east and south neighbours according to a 3-bit function specification. The genotype uses Cartesian Genetic Programming to represent the function of a cell and its update logic as indexed graphs encoded in linear strings of integers.

This model has been successfully applied to evolving a 2-bit multiplier (Liu, Miller, and Tyrrell 2005a) and evolving a 2-bit multiplier with fault tolerance to transient faults (Liu, Miller, and Tyrrell 2005b).

### 3.3.4   Computational Models of Gene Expression

There have been several computational models of gene expression developed, many of which have been used to study the dynamics of gene regulatory networks (GRNs), for example the work of (Reil 1999; J. and Osborn 2000). Models based on Random Boolean Networks (RBNs), proposed by Kauffman (Kauffman 1969), have particularly been applied to this domain. Gene expression models have been successfully applied to evolving deterministic finite-state automata (DFA) which induce the Tomita language set (Luke, Hamahashi, and Kitano 1999) and distributed fitness evaluation (Kargupta and Park 2001).

There has also been some work with models used to generate constructs based on the information expressed in the genome, by simulating some of the most important processes in gene expression. This type of approach has been applied to growing structures (Eggenberger 1997b; Kumar 2004), generating evolutionary building blocks (Wu and Garibay 2002), neural networks (mostly for robot control) (Jakobi 1995; Eggenberger 1996; Eggenberger 1997a), robot control (Bentley 2004) and evolvable hardware (Roggen, Floreano, and Mattiussi 2003).

Artificial models of gene expression tend to treat genes as rule-based systems, with a pre-condition that needs to be satisfied for the post-condition encoded in the gene's coding region to be performed. Generally, simplified gene-encoding models are used with genes either being of fixed length and placed sequentially on the chromosome (for example the approaches of (Eggenberger 1997b; Bentley 2003; Roggen, Floreano, and Mattiussi 2003; van Remortel, Manderick, and Lenaerts 2004)), or are identified with a short (typically at least four characters) standardised promoter sequence (see (Jakobi 1995; Reil 1999; J. and Osborn 2000) for examples of this). In both approaches genes have regulatory regions (encoding the gene's pre-conditions) directly upstream of the gene or promoter, which are used to determine whether the gene is expressed.

These models were influential in determining aspects of the gene expression model developed later in this book, and as such, some details of these are given below, before concluding this section with a few details of the gene expression models used in EHW.

#### Template Matching-based Models

Jakobi (Jakobi 1995) encoded genes on a chromosome comprised of 4 different symbols, and identified genes with a promoter sequence, equivalent to a "TATA" box, followed by a threshold region, link template and a coding region. The threshold region specifies how much stimulation is required to activate the gene, while the link template determines which proteins can affect the gene's activation. The coding region begins with a triplet defining the protein type (signal, mover, splitter, differentiator, dendritic, threshold) and is followed by a sequence of 64 characters, grouped into triplets, that code for the protein. Template matching is used for protein-genome and protein-protein interactions. This is done by forming the 64 characters of a protein string into a ring by joining the beginning and end of the protein string. In the case of the gene, the protein is rotated over the link template to find a match. When a match is found, its affinity is defined according to the weighting of the match between the diametrically opposite side of the protein string (to this match) and an arbitrary, but fixed template that could be viewed as being part of an RNA polymerase molecule.

In Reil's gene expression model (Reil 1999) a standard gene promoter of "0101" is used to define the start of genes on a chromosome comprised of a string of digits. The product of a gene (the protein) is given by increasing each digit of the fixed length gene (6 digits) by one. Regulation is achieved by the binding of gene products to matching sequences (template matching) of the gene regulatory areas directly upstream from the gene. Regulatory sequences can act as enhancers or inhibitors, depending on the value of the final digit, with inhibition having priority.

Eggenberger (Eggenberger 1996; Eggenberger 1997b) used a more complex chromosome model, in which the chromosome is comprised of a sequence of digits 0-6, of which the digits 1-4 are used to represent the regulatory and coding regions, while "0" is used to demarcate adjacent regulatory regions, "5" is used to indicate the end of a regulatory region and the beginning of a structural gene (a coding region), and "6" indicates the end of the coding region, and the beginning of the next regulatory region (for the following gene). Gene activations are determined by subtracting the value of a transcription factor (a string of digits from 1-4) from that of a regulatory region (typically the first 6-8 digits), with the result indicating the affinity of the transcription factor. Positive results act as gene activators, while negative results produce an inhibitory effect on the gene. The concentrations of every transcription factor at a gene's regulatory regions are calculated and the products summed, after which the result is put through a sigmoidal function. If a fixed threshold is exceeded then the gene is activated or inhibited. When a gene is activated it may produce a transcription factor for gene regulation, a cell adhesion molecule for building connections between neurons, a receptor for regulating communication between cells, or an artificial function such as cell division or cell death.

## An Operon-based Model

Kennedy and Osborn (J. and Osborn 2000) created a model of gene regulation based on the operon model (see Section 3.2.3.1 for more details on the operon model), complete with promoters and associated regulatory regions, genes and non-coding regions all identified by 4-bit sequences (aligned on 4-bit boundaries) on a binary chromosome. Each 4-bit sequence is decoded to a number from 0-15, with the first 10 digits representing monomers for constructing proteins, and the others representing *start operon* (10 and 11), *start enzyme* (12 and 13), *start carrier* (14) and *end operon* (15). Promoter regions may contain regulatory sequences that act on the following operon. Operons may be always active, active unless the promoter is disabled, or inactive unless the promoter is activated. The activation of the promoter in these cases is regulated by the presence of molecules in the cell that match the regulators key sequence. Molecules strength of binding, and hence duration of binding, is according to how many bases match between the molecule and the promoter. Regulation is performed at the transcription level, with "spiders" attaching to promoters and working its way along the operon transcribing its gene products and then detaching once the operon's end is reached and immediately reattaching randomly to another active operon promoter. Speed of transcription initiation is based on the spider's similarity in shape to that of a fixed ideal.

**Gene Expression Models in EHW**

Roggen et al. (Roggen, Floreano, and Mattiussi 2003) use a simple rule-based scheme where an expression table is encoded directly on the genome, with the function expressed by the gene (the corresponding spiking neuron's function) determined by a lookup of the protein signal intensities.

Bentley (Bentley 2003) uses a similar scheme, with genes having a precondition (a fractal protein shape), and a threshold (of similarity) for the gene (post-condition) to be expressed, resulting in the production of a fractal protein (with the shape being encoded in the gene in the same manner as in the pre-condition, as 3 real valued numbers).

The work of van Remortel et al. (van Remortel, Manderick, and Lenaerts 2004) also models genes as having a single pre-condition corresponding to a given property of the immediate environment (cell), neighbour, or the global development environment. Genes also have a single post-condition, which can be a change in protein concentration, a cell division, or cell differentiation. Genes are evaluated from left to right on the chromosome.

## 3.4    Summary

This chapter has covered the properties of morphogenesis that make it attractive for evolvable hardware. Specifically, morphogenesis provides scalability, which is essential for EHW on modern commercially available FPGAs, as well as being able to ensure that solutions are valid configurations. While the morphogenesis process itself offers a great deal of scalability, this is further enhanced through a high degree of redundancy in the genetic encoding, providing neutral pathways through evolutionary space and genetic robustness.

Furthermore, morphogenetic approaches also provide the opportunity to utilise the modular nature of the developmental processes for performing incremental evolution, though this will not be utilised in the work presented in this book.

From the review of morphogenesis in EHW and evolutionary computation, it appears that cell-based models driven by gene expression offer the most promise for evolving circuits within the constraints of an FPGA. Successful examples of this approach, though not on FPGAs, include the work of Jakobi in evolving robot controllers (Jakobi 1995) and the POEtic system (Roggen, Floreano, and Mattiussi 2003).

L-systems, on the other hand, while successful in generating complex structures in unconstrained environments, have several issues that make them unattractive for gate-level EHW on an FPGA. The issue of how to choose primitives and rule representations such that evolved constructs fit within the constraints of the FPGA architecture, is at best difficult and at worst, intractable. Such an approach is also likely to be fragile, as shown in the work of Haddow and Tufte (Haddow, Tufte, and van Remortel 2001) who applied L-system rules directly to the configuration bitstream of the virtual FPGA, with poor results, requiring the crossover operator to be removed to achieve some limited success.

Hence the cell-based developmental approach driven by gene expression is the one that will be taken for the remainder of this book. To derive a suitable model of this is the aim of the following chapter, whereby the biological processes and structures covered in Section 3.2 of this chapter, are evaluated for their applicability to EHW on an FPGA, and then based on this a model of morphogenesis for EHW on an FPGA can be constructed.

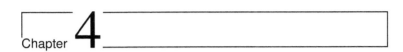

# Chapter 4

# Designing a Morphogenetic EHW System

## High-Level View of Morphogenesis Process

From previous chapters, it can be seen that, at a high level, the morphogenesis process can be seen as a mechanism for cellular regulation and coordination as specified by genes on the chromosome, and instantiated by proteins, which act both as building blocks and signals. This provides a view of the cell as a set of parallel processing elements (the polymerase that transcribe genes) controlled by the interactions between their programs (genes encoded in the chromosome) and their environment (as represented by proteins detectable by the cell).

This process is used to explore phenotype space, as directed by evolution, which in turn is guided in its exploration of genotype space by selection pressures based on the fitness of the phenotype in its environment (which in EHW is an externally defined problem posed to the population).

## Application of Morphogenesis Advantages to EHW

In this chapter a morphogenetic model for EHW is constructed based on this view, and keeping in mind the specific advantages of a morphogenetic approach that were outlined in Chapter 3.1, these being: scalability, genotype robustness, phenotype legality, neutral pathways, developmental modularity, and incremental evolution. Briefly, these advantages are provided by a developmental process for addressing scalability, while also providing the opportunity for modularity to be utilised (by the experimenter) through incremental evolution, and by using a genetic encoding for the developmental model along with supporting genetic operators, which provide genotype robustness, structurally legal phenotypes, and utilise neutral pathways to aid evolutionary search.

## Model Overview

The model that was chosen is comprised of a set of prokaryote-like cells, which contain a copy of the single chromosome, polymerase enzymes for transcription, and a number of proteins, all of which double in functionality as building blocks and transcription factors. Genes are comprised of a coding region, which may encode multiple proteins, a promoter, and enhancer and repressor regulatory regions. Genes are regulated at the transcription level, with transcription being performed by polymerase enzymes, and translation and protein construction being integrated in this process. Inter-cellular coordination is achieved using simplified signaling pathways between neighbouring cells.

The *morphogenetic evolvable hardware* (MGEHW) approach outlined in this chapter is based on the following premises:

- Developmental processes should be closely tied to, and utilise, the underlying architecture.

- Evolution should be able to learn useful relationships between components.

- Transcription-level gene expression is sufficient for learning and expressing these relationships.

## Model Breakdown

The first step in designing this model for EHW is to choose a useful subset of FPGA hardware resources, and then create a mapping between the FPGA hardware resources chosen and biological constructs. With the basic structures decided on, then the developmental processes and mechanisms that will be implemented can be decided. Then a gene expression model and its encoding on the chromosome can be devised to drive these. Lastly a set of genetic operators for evolution to manipulate the gene expression specification can be chosen. These aspects of are addressed in detail in the following sections.

# 4.1    Choosing Resources for EHW

To minimise the complexity of the initial model, IOBs and BRAM are not manipulated, and only a subset of the resources available to each CLB is used.

## CLB Slice Resources

The subset of the CLB slice internals that was chosen is the two function generators (LUTs) with their associated flip flops, inputs and outputs; clock, clock enable and the flip flop set-reset inputs; and the slice control mux that configures the LUTs to act as function generators, RAM or shift registers. These resources are illustrated in Figure 4.1.

Note that the carry in signals and carry logic are not used. The F5 (combining 2 LUTs into one 5 input function generator) and F6 (combining the 2 LUTs from the other slice with the 2 LUTs in this slice to produce a 6-input function generator) muxes are also not used.

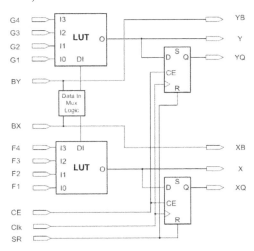

Figure 4.1: MGEHW System Supported Slice Resources

## CLB Routing Resources

A subset of routing resources was also decided on, again to minimise the complexity of the initial EHW model. In this case it was decided that only local to CLB and direct CLB to CLB single lines would be used. Hexlines, long lines, single-to-single lines (used for turning corners, for example), and the tristate bus are not used. These limitations provide a view of the CLB as illustrated in Figure 4.2.

## CLB Resource Summary

Together this gives 2 x 13 slice input muxes, 4 x function generators, 2 x 1 slice configuration mux (shown as "Data In Mux Logic" in Fig. 4.1), 2 x 6 slice outputs, a single 8 line CLB output bus, and a CLB output bus to single routing mux.

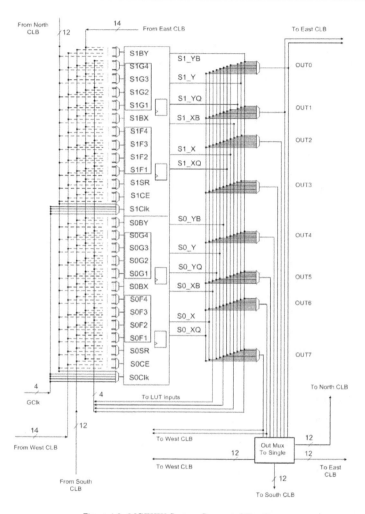

Figure 4.2: MGEHW System Supported Routing

## 4.2    Mapping Biology to Hardware Structures

### Degree of Hardware-Biology Correspondence

There is a spectrum of approaches as to how to map from a cellular model to
the underlying hardware of the FPGA. On one extreme it may be a totally

simulated cellular model, with no correspondance between the components and processes of development and the underlying hardware, with only the result of development being implemented on the FPGA. To the other extreme where all aspects of development, such as proteins, signal pathways, etc., correspond to actual physical resources on the FPGA.

## Development Closely Tied to FPGA Architecture

After an in-depth look at both the Virtex architecture (see Section 2.4 on page 33 for details) and biological developmental processes, a model in which the developmental process is closely tied to the FPGA structure was chosen. Rather than trying to evolve or design simulated developmental mechanisms (such as special signaling proteins and signal receptors, etc.) and structures (like cells and proteins), the underlying FPGA structure is used for implementing much of the developmental process. Cells are mapped to fixed FPGA regions, proteins within a cell correspond to the FPGA resources within the cell's region, all proteins are transcription factors and hence able to affect gene expression, and inter-cell signaling is done using shared FPGA routing resources.

## Advantage over Simulated Approach

This approach counters some of the difficulties involved in having simulated developmental processes distinct from the underlying medium. Problems such as the computational expense and, in particular, the arbitrariness of various aspects of development, such as how to map from cell state to FPGA state, determining which proteins are signals, rates of diffusion of signal proteins, what their extent of spread is, and matching them to receptors, which themselves need to be designed somewhat arbitrarily or evolved.

As the evolutionary and developmental process is tied closely to the underlying hardware medium, evolution is able to manipulate the FPGA structure directly at the level of multiplexor settings, and growth is implicitly derived from the evolved relationships between hardware settings regulated through the expression of genes.

## Cell Mapping

Cells are mapped directly to the underlying FPGA architecture, so that each cell may correspond to either a CLB slice or logic element (a LUT and its associated flip-flop), according to the granularity desired by the user. The decision to map to slice or logic element, rather than CLBs, was made due to the relative functional independence of these elements in the Virtex FPGA. While routing resources may either be dedicated or shared across the CLB, LUTs act either as independent 4-input Boolean function generators or LUT pairs that can be combined within a slice to form a single 5-input function generator. For sequential logic, the two logic elements (LEs) within a slice share clock and control signals.

## Protein Mapping

Proteins are the basic building blocks of biological systems, and are also used for controlling gene expression. In the morphogenetic EHW model developed in this work, proteins represent the FPGA's gate-level resources accessible through JBits, thus allowing cell behaviour to be controlled by the state of the underlying logic element or slice.

In this approach, each FPGA resource is represented by a single unique protein, which will have several possible setting values possible (for example *ON* or *OFF*), of which only one may be active at any given time. Thus a multiplexor may have its setting changed when the next instance of the same protein is produced (for instance switching between on and off states), but there can never be more than one instance of that given multiplexor in the cell. This also means that each unique protein will always have an instance present in the cell, if that resource exists in the underlying slice or logic element.

## Mapping Signaling Pathways

For cells to coordinate their behaviour, it is necessary that some form of inter-cellular communication is available. In nature, this is achieved with signaling pathways, whereby a cell on receiving a signal from another cell causes the re-lease of a protein, which can in turn affect gene expression in the cell. This can be implemented directly in a morphogenetic EHW scheme by using the state of routing resources shared between cells (i.e. a line that connects from one to the other) to transmit directed inter-cellular signals (instantaneously). Generally this involves the signal receiving cell querying the state of the routing multiplexor in the signal emitting cell, however note that there is no require-ment for the developmental protein signal direction flow to correspond to the electronic signal flow.

## Example of Bio-Hardware Correspondence

See Figure 4.3 for an example of the correspondence between cells, proteins, and FPGA configuration. Cell-local proteins are shown as circles, and the signaling proteins between the two cells are shown as stars. Note, however, that only the proteins that are discussed here are shown. In the West cell (located in CLB Y,X), out bus line 0 is driven by the registered output of the function generator (represented by protein *OUT0.S0_YQ*), and out bus line 7 drives single line East 20 (locally represented as *OUT7_To_Single_East20, Out-MuxToSingle.ON*). As this latter resource connects to the cell in the adjacent CLB to the East (CLB Y,X+1), this easterly cell is able to detect the state of this resource (ON or OFF), as indicated by the presence of a corresponding signaling protein. This cell also has a local protein that indicates that LUT input line 4 is driven by single line West 5 (represented as *S1F4.Single_West5*). The cell to the West is able to query the state of this LUT input, and so has a corresponding signaling protein present. Note that for each possible line to this LUT input, the originating cell is able to test whether or not the LUT input is being driven by this particular line, but is not, however, able to query if the LUT input is driven by a line originating in another cell. The same principle applies to querying cell output lines.

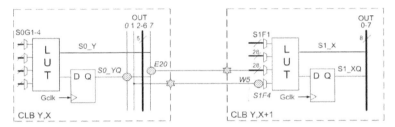

Figure 4.3: Example Cell/Protein to Hardware Correspondence

## 4.3    Developmental Processes

Biologically speaking, development is the process by which a multicellular organism is formed from an initial single cell. Starting from a single cell, a simple embryo is formed comprised of a few cell types organised in a crude pattern, which is gradually refined to generate a complex organism comprised of many cell types organised in a detailed manner. This model of the development process is known as epigenesis, and, as mentioned before, is comprised of five major overlapping processes: growth, cell division, differentiation, pattern formation and morphogenesis. The applicability of each of these processes to FPGA-based EHW is covered below.

### 4.3.1    Growth and Cell Division

Although growth and cell division are essential in biological organisms, their applicability to an EHW model is limited by the fixed mapping from cells to hardware that has been chosen. The underlying FPGA hardware has a fixed structure, with FPGA logic cells (CLBs) being fixed in both shape and in their physical relationship to each other (having a regular matrix pattern), with only their connectivity and function being variable.

### 4.3.2    Pattern Formation and Differentiation

While pattern formation and cellular differentiation involving cell-division are not directly applicable due to the fixed cell-hardware mapping, there are some mechanisms that may still be relevant, such as morphogens, cytoplasmic determinants and axis specification. These require the use of simulated molecules, however, and as such will be covered later with other secondary developmental mechanisms in Chapter 7.

Cellular differentiation involving changes to cell function (rather than cell structure) as caused by external signals from other cells, is known as induction, and is relevant and important, however. This form of cellular differentiation is indistinguishable from the inter-cellular coordination of morphogenesis in the MGEHW model.

Inductive signals may occur over various ranges and may produce a single standard response in the responding cell, or a graded response dependent on signal concentration, in which case it is called a morphogen (Twyman 2001). Induction and other forms of signaling (both from within and without the cell) can be readily applied to EHW with fixed cell structures, and along with gene expression, are probably the most important mechanisms of developmental biology in their applicability to EHW.

### 4.3.3    Morphogenesis

The process most useful to an EHW model with a fixed mapping between cells and hardware is morphogenesis, through which cells coordinate their behaviour, and this is driven by gene expression, which will be covered later in Section 4.4. Morphogenesis is the process by which complex structures, such as tissues and organs, are generated through the utilisation of cell behaviours.

Obviously many of the biological cell behaviours, such as changes of cell shape and size, cell fusion and cell death, adherence and dispersion, movements relative to each other, and differential rates of cell proliferation are not directly applicable to developmental processes in EHW where there is a fixed mapping between cells and the underlying hardware structure. Cell behaviours here are limited to changes in connectivity and function, hence the only biological behaviours applicable to the MGEHW approach are cell-cell and cell-extracellular matrix interactions, and cell death.

**Cell-Cell Interactions**

Central to cell-cell interactions are signaling pathways, by which protein signals are released in the receiving cell, which are in turn able to affect gene expression. The mechanics of signaling pathways are quite complex, and not necessary for EHW. What is necessary is that signals from other cells can be detected and affect the expression of genes within the receiving cell. For this reason, and due to the direct correspondence between proteins and FPGA resources, cell-cell interactions are limited to interactions between cells that share some routing resource, such as single-length lines between cells in adjacent CLBs [1], or shared out bus lines for cells within a CLB. Hence signals are simply the (querying of the) configuration of the shared resource.

**Extracellular Matrix**

The notion of an extracellular matrix, a network of macromolecules secreted by cells into their local environment, could have relevance to a morphogenetic EHW system if inter-CLB routing resources are used, specifically the programmable interconnection points (PIPs) used to connect lines from other CLBs to lines that can connect to the local CLB's inputs or outputs. However, as only directly connectable single-length lines are used in the current implementation, the need for an extracellular matrix is eliminated.

**Cell Death**

Cell death is another behaviour that could be useful in a fixed cell-hardware mapping. Although it is not currently provided, it would be simple to implement, by disabling connections to and from the dead cell, and could be used to isolate faulty regions of the underlying hardware.

# 4.4    Gene Expression Model

## Gene Expression Drives Development

The whole biological developmental process is largely driven and controlled by the mechanics of gene expression. This is the process by which proteins signal cellular state to activate genes encoded on the chromosome, which in turn are able to effect changes in cell state by the production of further proteins which may be used as signals or cellular building blocks. This paradigm is also applied to the morphogenesis model.

---

[1]Note that in the current version of the MGEHW system, the only inter-CLB routing provided is directly connectable single-length lines between neighbouring CLBs.

## Transcription Level Gene Expression

Gene regulation at the transcription level appears to be the most important level of gene regulation, furthermore, the results achieved by Reil (Reil 1999) who used a gene expression model with transcriptional regulation, demonstrated that gene regulation using a simple model is able to produce many of the properties exhibited in nature. For these reasons, and to limit computational expense, gene expression is regulated solely at this level in the MGEHW model.

## Transcription Regulation Model

The model of regulation of gene expression that was developed is one in which the expression of a gene is based on the attraction of polymerase to a given gene's promoter, ready for gene transcription to take place. This is influenced by the binding of proteins at regulatory regions. Proteins binding on the enhancer, upstream of the promoter, increase the likelyhood of attracting polymerase to the promoter for transcription to take place, while binding of proteins on the repressor, lying between the promoter and the gene coding region, effectively block polymerase, thus decreasing the likelihood of transcription occurring. Once transcription begins, however, the binding of proteins at the regulatory regions has no effect, as further transcription is blocked until the gene's transcription has completed. Furthermore, transcription rate is constant, and not dependent on binding at regulatory regions, or on the promoter's signature (there is only a single promoter signature used in this model, which is covered in Section 4.4.1).

## Gene Transcription

From a biological viewpoint, for a gene to be expressed, it must be transcribed by polymerase to RNA, which is then translated by ribosome, according to the genetic code, to form a chain of amino acids, which then fold into a protein structure.

For EHW, a similar model can be used, whereby polymerase moves along the coding region (transcription) generating an intermediate string, analogous to a chain of amino acids, specifying the resource and its setting according to the genetic code (translation). A "protein" can then be formed (protein folding) from this intermediate format by decoding the fully specified FPGA resource into JBits class constants and settings values for manipulating the FPGA configuration bitstream.

While there is corresponding processes in the MGEHW model for each of the biological processes of transcription, translation and protein folding, these were not chosen to emulate biological artefacts, instead these analogues are implemented for practical purposes. Transcription provides a temporal dimension to gene expression; translation to an intermediate format (FPGA resource, attribute, setting) provides flexibility and redundancy of encoding; "protein" folding allows the amenable intermediate format to be mapped to valid FPGA configuration settings (JBits classes and constants).

## Template-Based Protein Binding

In the MGEHW model all proteins are treated as transcription factors, so that choosing which effect gene expression can be decided by evolution (via elements that can bind at binding sites) rather than arbitrarily by the designer. Binding is determined using a template model, whereby a protein is able to bind to a chromosome section, if the protein's signature sequence matches a complementary sequence of bases (0 and 2 bind as do 1 and 3) on the reversed and folded chromosome (covered later in Section 4.4.1).

Template models have been successfully applied in the work of (Reil 1999) and (Jakobi 1995). While this approach provides unlimited interactions between the chromosome and proteins, and proteins themselves, it can be slow due to string matching. However this can be resolved by preprocessing the genome (as was done in (Jakobi 1995)).

## Protein-Chromosome and Protein-Protein Interactions

For the MGEHW system, a model has been chosen in which proteins are only able to bind at specific bind sites within regulatory regions, with template matching determining binding. No protein-protein interactions are provided for, as this would introduce unnecessary complications to the model. This provides a simple, but effective model in which protein-chromosome binding provides a means of generating cell state transitions, with all interactions mediated by gene expression, without explicit interactions required between proteins in the cell. However, these could be seen as implicitly occurring via gene expression, whereby one or more proteins would be able to affect other proteins by regulating the production of proteins.

## 4.4.1    Chromosome Encoding

### Base-4 Encoding

While most genetic algorithms use a binary encoding on a fixed length chromosome, a base-4 chromosome was chosen so as to provide two distinct base pairings (0 and 2, and 1 and 3) for sequence binding (in a similar manner to the manner in which the two complementary strands of DNA are bound together), and, to give more redundancy for neutral mutations in the genetic code. Furthermore, this allows the base transitions that are provided by the base mutation operators to be constrained.

### Variable-Length Chromosome

To provide a model that is as flexible as possible, a variable length chromosome has been used with genes and their regulatory regions being identified by special signature sequences rather than by location. This means that no prior knowledge is required regarding the number of genes, or the specifics of their pre-conditions, needed to solve a problem. Furthermore, the unused regions of the chromosome that don't participate in gene coding or control, often referred to as 'junk' DNA, may act as a scratch pad for evolution.

### Gene Structure

Genes are comprised of a coding region, from which proteins are generated, and generally, a promoter sequence upstream of the coding region to which polymerase is attracted, with regulatory regions adjacent to the promoter that influence the initiation of gene transcription. The regulatory regions (the enhancer upstream of the promoter, and the repressor located between promoter and coding region) are comprised of a variable number of bind sites to which proteins are able to bind, with the regions between being unused (i.e. junk DNA). Note that in this design it is possible to have genes without a promoter, with the corresponding lack of associated regulatory regions, however these genes may be ignored by the morphogenesis system implementation (as is currently the case) or assigned a low transcription rate (for example).

### Gene Encoding

Genes are identified on the chromosome by the presence of coding regions, demarked by a start codon, followed by a set of resource(s) encoding codons, and a stop codon. If there is a promoter sequence (0202, analogous to the biological TATA sequence) upstream of the coding region, but downstream of the upstream gene's coding region, then the enhancer and repressor regions are identified adjacent to this, extending all the way to their neighbouring gene coding regions.

Within the regulatory regions, sequences are reversed, folded and interleaved (to prevent problems with bind signatures that may otherwise be mistaken as a coding region), to generate a sequence of bind codons, that can be decoded to locate bind sites, identified by bind start and stop bind-codons, and the binding resource(s) signatures.

### Example Chromosome Section

Figure 4.4 illustrates gene structure and encoding with an example chromosome section. This shows how multiple genes are encoded on the chromosome, and highlights one of these genes with its promoter, coding region, regulatory regions and bind sites (denoted in the figure as 'bs'). The repressor is shown in greater detail, including the three sets of codons that are extracted, and provides a graphical view of how the bind site is located within this region.

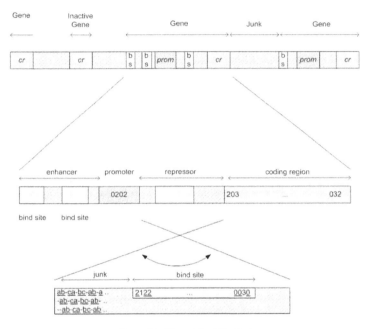

Figure 4.4: Chromosome Regions

## 4.4.2    Codon-based Genetic and Protein Bind Signature Codes

### Neutral Mutations

A codon-based genetic code was decided on to facilitate neutral mutations on gene coding regions: most single base mutations will result in either no change or a change to a related gene product, especially for mutations in the third base of a codon, and for mutations from one base to its complement. For the same reasons, a related codon-based bind signature code is also used to determine binding between proteins and bind sites.

### For Encoding Growth

This genetic code was specifically designed for use with EHW systems where the number of resources to be set per CLB is not predetermined, such as when encoding a growth process. One start and three stop codons, analogous to those found in nature, are allocated for delimiting gene coding and regulatory bind sites. These regions encode FPGA gate-level logic resources (and simulated molecules if they are used), and allows multiple of these to be encoded per gene.

### Gene Products

The other 60 codons are allocated to encoding gene products (FPGA muxes or simulated resources), but unlike nature where each codon completely specifies a distinct amino acid, in this scheme several codons are required to encode a single gene product (or, in the bind code, to encode the resource signature). The first codon of a gene product specifies its resource type, and may also (fully or partially) specify the resource attribute. The following codons then specify the resource's setting (and if necessary, also the rest of the attribute). The general format for a gene product is *resource, attribute, setting*. This format is further decoded (analogous to protein folding) to produce JBits class constants and settings values for manipulating the FPGA configuration.

### Resource Types

From Section 4.1 it can be seen that there are five main FPGA resource categories available in the CLB: slice inputs, LUT function, slice configuration (LUTs as function generator(s), shifter, RAM), slice output to CLB out bus, and out bus to single line PIPs. These groups of resources correspond to the FPGA resource types used in the *resource, attribute, setting* format employed by the genetic and bind codes. The corresponding resource types being *SliceIn*, *LUTBitFn* (and other LUT encodings), *SliceRAM, SliceToOut*, and *OutToSingle\** ( * = *Bus* for singles organised by out bus line, or *Dir* for singles organised by direction).

### Resource Attributes

Within each of these resource types there are many actual FPGA resources, and so each resource type is combined with an attribute to identify the specific FPGA resource. For example *LUTBitFn,F*. Attributes may be comprised of more than one field, as is the case with slice inputs (where inputs are divided into those for the LUT and those for the control muxes); hence, for example *SliceIn, Ctrl, Clk*.

**Resource Settings**

For each configurable FPGA resource, identified by a resource and attribute, there will be a number of possible settings to which it can be configured. In some cases (out mux to single line PIPs for example) this is simply *OFF* or *ON*, while in other cases (slice input muxes) there can be more than twenty possible settings.

**Differences Between Bind and Genetic Codes**

The main difference between the bind code and genetic code is the addition of connecting types, for inter-CLB routing resources, to implement inter-cellular signaling pathways.

A prepended "_connect_" in the attribute field is used to indicate that this resource is a connecting resource, which will have the state of the mux in the connecting CLB queried to determine the protein representing the signaling pathway. Otherwise, the state of the resource in the current cell is queried to determine the local protein representing this.

The resource types that can be used for inter-cellular signaling are: SliceIn with a connect attribute, which will query the corresponding OutMuxToSingle mux in the originating neighbour of the connecting input line; OutToSingleBus and OutToSingleDir with a connect attribute, which will query the corresponding line's slice input muxes in the connecting output line's destination neighbour; SliceToOut with shared CLB out bus lines is able to influence the other cells within the CLB; SliceIn with recurrent lines, from within the CLB, is able to be treated as a signal between cells that share the CLB.

Another difference, in the current implementation, was the allocation of simulated molecule (TF) codons as markers for bind sites (TFs are bound according to their bind sequence and distance from source, rather than according to their resource, attribute and setting).

Further details of the genetic and bind codes, including codon allocation, are given in Appendix C.

## 4.4.3   Example of Gene Expression Driven by FPGA Configuration

As is the case in biological gene expression, in the MGEHW model genes encode proteins, which represent the configuration of a given FPGA resource in the cell (slice or logic element) and these same proteins also regulate the transcription of genes. In this model, any protein may act as a transcription factor, binding to the regulatory regions of genes (note proteins can bind to many sites concurrently). Each protein in the following examples maps directly to a single FPGA resource. A gene is activated by its enhancer being bound to by a protein, representing the current FPGA resource state. If the gene's activation is strong enough to attract free polymerase, then the gene is transcribed to produce proteins, which configure the cell's FPGA resources.

Figure 4.5 shows an example of a gene being activated by the current state of the FPGA resources underlying the cell. Then when transcription takes place, one of the two (re)configured FPGA resources binds to the repressor site of this gene, disabling further transcription. For clarity, the JBits and intermediate representations are shown in this figure, along with the chromosome encoding, and associated hardware resources.

A step-by-step break-down of the sequence of events displayed here is given in the following set of figures (Figures 4.6–4.10). Note that this is purely a graphical representation of concept, and is not meant to indicate the actual inner workings of an implementation of the morphogenetic model.

In Figure 4.6 polymerase is attracted to the promoter site by the binding of a protein on the enhancer bind site. This protein, which corresponds to the *S0G1* LUT input mux being set to *OFF*, binds (according to its bind code) to a complementary interleaved sequence on the enhancer bind site. Binding of an FPGA resource to a bind site involves querying the resource associated with a given subsequence of the bind site, to determine its setting. If the current setting matches that encoded on the sequence from the bind site, then the protein associated with that resource is considered to be bound.

Once polymerase has been attracted to the promoter, gene transcription begins, starting at the start codon (*203*) in the coding region. Polymerase works its way along the coding region, a given number of bases per time step, using a genetic code to translate from codons to an intermediate form, which is further decoded to the JBits format, resulting in the release of a protein in the cell, and the configuration of the associated FPGA resource. This process is shown in Figure 4.7.

In Figure 4.8, the protein produced in the previous step has bound to the repressor bind site, and another protein has been transcribed and its associated FPGA resource configured.

When the polymerase reaches a stop codon (here *032*), it is released from the chromosome, terminating transcription of the gene, as shown in Figure 4.9. Once the polymerase has detached from the chromosome, it can then be attracted to the promoter site of this or another gene on the chromosome.

Figure 4.10, shows how the binding of a protein to the repressor blocks further transcription of the gene from occurring by blocking polymerase from binding to the promoter.

This is, however, an over-simplification to illustrate the concept; in reality the actual MGEHW implementation uses the scheme presented in Section 5.4, where the determination of a gene's expression is based on the strength of activation according to the relative strengths of enhancer and repressor activation according to their bind sites, and the relative strength of the gene's activation in relation to other genes not currently being transcribed.

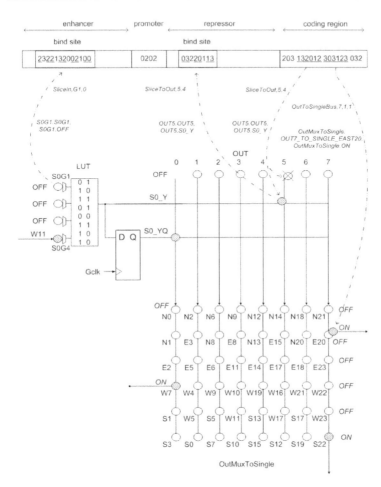

Figure 4.5: Example of Gene Expression Driven By Hardware State

## 4.5   Genetic Operators

For evolution to explore genotype space, genetic operators are required for combining (generally fit) parents' chromosomes to generate offspring in the form of a new chromosome. Several biologically inspired genetic operators

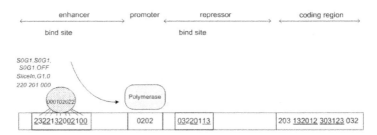

Figure 4.6: Initiation of Transcription with Polymerase Binding

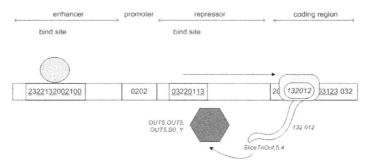

Figure 4.7: Gene Transcription

have been devised to work with variable length chromosomes and redundant genetic and bind codes, and to facilitate the generation of new chromosomes with a minimum of disruptive effects to the phenotype. These operators are homologous crossover, inversion, base mutations and frameshift mutations.

### 4.5.1    Homologous Crossover

In nature crossover requires two DNA molecules with a large region of homology (nearly identical sequence), usually hundreds of base pairs long, so that the exchange between two chromosomes is usually conservative (Winter, Hickey, and Fletcher 2002). This prevents biological crossover from having the disruptive effects associated with crossover in genetic algorithms.

Due to the use of variable length chromosomes, a specialised crossover operator, known as homologous crossover, has been implemented which, taking inspiration from nature, performs crossover at a point on the parent chromosomes with a large degree of homology. This is done using a variant of

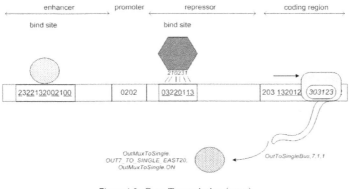

Figure 4.8: Gene Transcription (cont.)

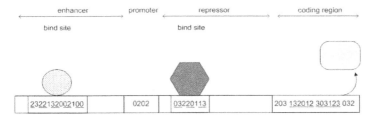

Figure 4.9: Gene Transcription Terminates

the longest common substring algorithm, implemented using Ukkonen's algorithm for constructing suffix trees in linear time (Ukkonen 1995). However,
rather than choosing the longest common substring, a common substring is
randomly chosen biased towards longer substrings. One-point crossover is
then performed at the boundary of the randomly chosen subsequence. This
form of crossover, along with the signature-based encoding of the chromosome,
provides a non-disruptive method of combining genetic information from two
parent chromosomes.

An example of homologous crossover is shown in Figure 4.11.

## 4.5.2   Base Mutation

Mutation in this system is also biologically inspired, to take advantage of the
redundancy in the genetic and bind codes, such that many of these mutations will have no effect on the encoded protein or bind signature, and thus

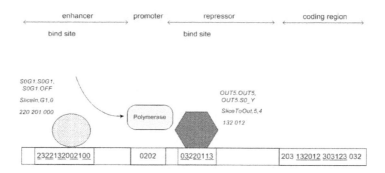

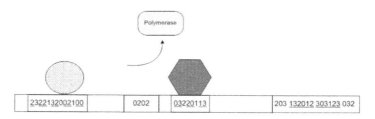

Figure 4.10: Gene Transcription Repressed

aid in evolutionary search. Mutations of a single base may be of two kinds: transversions, which involve flipping the base to its complement (0-2 or 1-3) and transitions, whereby a base is swapped with its non-complement (0-1 or 2-3). This is illustrated in Figure 4.12.

### 4.5.3    Frame Shift and Inversion Mutations

Other mutations may involve the insertion or deletion of bases in a sequence (illustrated in Figure 4.13), which may cause a frame shift in the codons downstream (in a coding region), and would thus have a serious effect on the encoded protein, which is generally deleterious in biological chromosomes (Winter, Hickey, and Fletcher 2002).

Another type of mutation, known as inversion, involves the reversal of a section of the chromosome, and is illustrated in Figure 4.14.

These forms of mutation are provided for completeness, though prior to

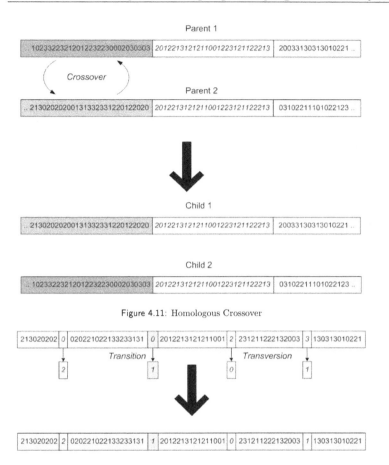

Figure 4.11: Homologous Crossover

Figure 4.12: Single Base Mutations

experimentation it was unknown what their utility would be.

## 4.6   Summary

In this chapter, a conceptual model of a morphogenetic evolvable hardware system has been formulated. This has been done by choosing a close hardware-

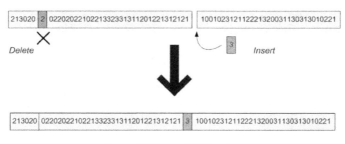

Figure 4.13: Frameshift Mutations

Figure 4.14: Inversion

biological model mapping to avoid unnecessary complications and arbitrary design decisions. EHW cells are mapped to CLB slices or logic elements due to their functional independence. Within each cell exist a set of proteins, each protein corresponding to an FPGA multiplexor and its current setting.

The biological mechanisms and processes were identified that have shown their utility in generating complex organisms. Only those that can be readily applied to the construction of circuits on a fixed size CLB region of an FPGA were chosen. Hence morphogenesis driven by gene expression was chosen as the primary biological processes applicable to this EHW model.

Transcription level gene regulation has shown its worth in evolutionary computation and such a model, loosely based on that of prokaryotes, has been decided on. Gene transcription is carried out by polymerase, which is then translated using a redundant genetic code to a flexible intermediate format before finally being "folded" into the JBits class and values required to access the FPGA resource. All proteins in the cell are able to bind to regulatory regions of the gene, according to the matching of their bind signature to its complement on the reversed and folded chromosome region, to modulate gene transcription by attracting or blocking polymerase enzymes from binding to the promoter.

To provide redundancy, a base-4 encoding was decided on, along with codon-based genetic and resource signature bind codes. A variable length chromosome with genes and their components identified by signature provides flexibility and redundant "junk" regions.

Lastly, genetic operators were chosen for supporting this model. This includes a biologically inspired homologous crossover, which performs crossover at regions of high similarity, thus reducing the likelihood of genetic damage. Several mutations operators were also chosen, with standard base mutations being designed to take advantage of redundancy in the genetic and bind codes.

In the following chapter the implementation of this model is detailed.

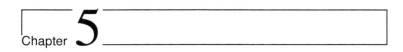

Chapter **5**

# Morphogenetic EHW Implementation

Using the design outlined in Chapter 4, a working morphogenetic EHW (MGEHW) system was implemented. This MGEHW system contains all the components necessary to seed, evolve, grow and evaluate digital circuits on a Xilinx Virtex FPGA. In this chapter, details of the implementation are discussed. For clarity this is divided into the major components of the system's process flow. This process flow is illustrated in Figure 5.1, noting that the use of simulated cytoplasmic determinants is optional, and will be covered later in Chapter 7.

## System Breakdown

From Figure 5.1 it can be seen that there are 5 major processing components, plus the interface to the FPGA itself. Briefly these six components are as follows:

**Setup** is responsible for generating and seeding the initial population of chromosomes, and for creating the FPGA and cell layouts and initial states.

**Evolution** applies the genetic operators to fit chromosomes to generate new members of the population, removes unfit chromosomes, and generates population statistics.

**Chromosome Preprocessing** extracts all the information required by the morphogenesis process from the chromosome, in the form of decoded genes with their associated promoter and bind sites.

**Morphogenesis** uses the information encoded in genes to direct the development of a circuit on the FPGA.

**Hardware Morphogenesis Interface** a JBits API based interface is used to update circuit configuration based on the developmental process.

105

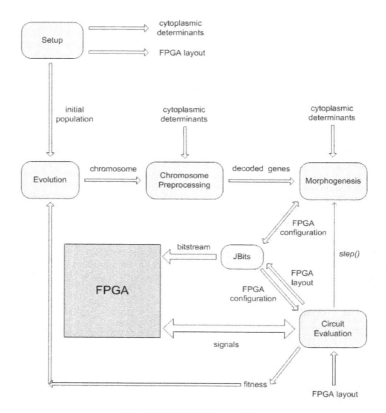

Figure 5.1: MGEHW Process Flow

**Circuit Evaluation** is performed at each stage of the circuit's development, and a fitness is assigned which is used to determine continuation of development, and for guiding evolution.

For a given run of the MGEHW system, setup is executed first after which evolution commences. During evolution, each generated chromosome is preprocessed and the decoded chromosome is passed to the morphogenesis process, which is instantiated and iterated through by the circuit evaluation process, to direct circuit development through the interaction between the decoded genes and the FPGA via a JBits hardware interface. The circuit evaluation process

106

returns a fitness metric for the chromosome to the evolutionary process, which is then able to continue until it completes according to the run's parameters.

Each these system components are elaborated in the following sections.

# 5.1    Setup

Prior to evolution and morphogenesis commencing there is a setup stage. This involves FPGA layout, preplacement of simulated molecules (if used), generation of an initial population and pre-evolution of this population to seed it with chromosomes containing a number of genes.

## FPGA Layout

The first task is to generate a layout for the Virtex FPGA, in the form of a bitstream, according to the designated CLB matrix size and IO specifications. In general the evolvable region's layout is setup so that the input bus signals are provided to single lines that can be fed to the LUT inputs around the center of the West edge, while the output bus signals are sampled from slice 0 G LUT and slice 1 F LUT based logic element (denoted here as S0G-Y and S1F-X, respectively) outputs (to ensure that the layout can also work with the limited CLB resources used in the first set of experiments) evenly spread along the East edge. There are several input single lines, from the West, available to the evolving circuit's inputs, while the outputs from the circuit are required to be routed to specific single lines to the East (East3 for S0G-Y and East11 for S1F-X logic elements). Signals are routed directly to and from IOBs on the perimeter of the FPGA to the CLBs adjacent to the evolvable region. This approach is an artefact of the original design aims, which were to allow the evolution of asynchronous circuits for robot control.

Also in this setup stage, if simulated cytoplasmic determinants are used, they are generated and pre-placed on the cell matrix. Pre-placement can also be used to preset elements of the FPGA to some desired initial configuration, by seeding cells with specific proteins.

## Initial Population Generation

Finally, an initial population of variable length chromosomes is generated, and then evolved for a given number of generations to produce "viable" chromosomes that have functional genes and regulatory regions, and neither too much or too little "junk" regions. Typically 25 generations was found to be sufficient to produce a diverse range of chromosomes that are functional, without converging the population prematurely on chromosomes that fit the designer's notions of viable. This approach saves unnecessary morphogenetic evaluation of large number of non-functional chromosomes in early generations.

### Chromosome Viability Fitness Measure

Viable chromosome fitness is dependent on 2 factors. The first factor is the number of genes in the chromosome, for which a Gaussian function is used, with a fast rise from 20-70 genes and a plateau around 100, scaled so that the plateau occurs at the gene ceiling (a problem-dependent empirically defined approximate value, in the experiments conducted to date a value of 8 was used). The second factor is the percentage of the chromosome that is deemed 'useful', for which a Gaussian function reflected about the 50% mark is used. This gives a rise from 0 in the range 0-30%, plateaus between 30-70%, and then falls from 70-100%. This ensures that the chromosome doesn't contain too much useless space, while also preventing the chromosome from being so tightly packed that evolution will have difficulty experimenting. The fitness can then be calculated as follows.

$$
\begin{aligned}
F \;=\; & (1 - \exp(-0.0005 * (\frac{100}{L} * N)^2)) \\
* \;\; & (1 - \exp(-0.004 * (50 - |50 - U|)^2))
\end{aligned}
\tag{5.1}
$$

The useful length of a given gene is given by

$$
\begin{aligned}
UG(i) \;=\; & P + \min(G_i, G_{max}) \\
& + \min(E_i, B_{max}) + \min(R_i, B_{max})
\end{aligned}
\tag{5.2}
$$

where $P$ is the length of the promoter ("0202"), which is 4 bases; $G_i$ is the length of the gene's coding region, including the start and stop codons, and $G_{max}$ is defined to be the length of the longest fixed-length gene product ($LUT$-$BitFn$ which defines the encoding of a LUT function in binary), which is 15 bases long, plus the start and stop codons (6 bases), giving $G_{max} = 21$; the length of the gene's enhancer ($E_i$) is the distance between the end of the previous gene's coding region and the start of this gene's promoter sequence, while that of the repressor ($R_i$), is the distance between the end of the promoter and the start of this gene's coding region, and $B_{max}$ is defined to be the length of the longest fixed-length bindable resource ($LUTBitFN$) plus bind site markers (6 bases) multiplied by $\frac{4}{3}$ to deal with the way regulatory regions are encoded, whereby each 3rd base is discarded, requiring 4 bases to encode 3, giving $B_{max} = 28$.

## 5.2     Evolution

This process is used to generate, monitor, and evolve a population of base 4 chromosomes, using a stable state genetic algorithm with tournament selection and no replacement. The genetic algorithm uses the biologically inspired operators introduced in Section 4.5, these being homologous crossover, mutation (transversions and transitions), inversion, and base insertion/deletion.

## Tournament Selection and Elimination

At each generation a breeding tournament is conducted to determine which chromosomes will be mated. An elimination tournament is also conducted, the losers of which are removed from the population, to be replaced with the offspring of the winners of the breeding tournament.

## Generating Offspring

Child chromosomes are produced by firstly applying homologous crossover to the parent chromosomes, the resulting two chromosomes then have inversion applied, followed by base mutation, which is either a transversion or transition, and base insertion/deletion. These genetic operators are each applied with a given probability. Specifically, the rates for crossover and inversion refer the probability of the operation itself being applied to any particular reproducing chromosomes, while base mutation and insertion/deletion rates refer to the probability of a base on a chromosome being modified (all child chromosomes have both mutation and insertion/deletion operators applied).

## Offspring Evaluation

The resulting offspring are then evaluated, and depending on the replacement option chosen, either both are placed in the population, or solely the fittest, while the other is discarded. Chromosome evaluation involves the other MGEHW system components for parsing and decoding the chromosome, performing morphogenesis, and evaluating circuit fitness. Once the fitness of the chromosome has been evaluated, an anti-gene-bloat mechanism may be applied to limit unbounded gene counts and chromosome growth.

## Unbounded Chromosome Growth

Due to the use of variable length chromosomes, it is possible for chromosomes to grow without bound. To prevent this unnecessary growth is penalised, which is measured in terms of the number of genes on the chromosome. This approach is used as chromosome length is tied to the number of genes encoded on the chromosome, and due to the fact that it is actually the number of genes that is the limiting factor in the MGEHW implementation: as the number of genes increases, so does the time required to update each cell. In biology chromosome growth is constrained by cell space, whereas in the MGEHW model chromosome growth is effectively constrained by cell processing time.

### Anti Gene Bloat Mechanism

The anti-gene-bloat mechanism is implemented as part of the fitness evaluation function as follows. After a chromosome in the population has been evaluated, if it is not the fittest chromosome evaluated so far and its number of genes ($G_c$) is greater than the rounded down mean of the best chromosome's gene count ($G_{max}$) and the problem specific gene ceiling ($G_{ceil}$), then its fitness ($F_c$) is scaled down by the ratio by which the number of genes exceeds this mean; i.e.

$$F_c = F_c * \frac{\text{mean}(G_{max}, G_{ceil})}{G_c} \tag{5.3}$$

In practice, the anti-gene-bloat mechanism has been found to be unnecessary, due to the low polymerase to gene ratio and high bind threshold that have produced the best experimental results. This probably prevents chromosomes with high gene counts having much advantage, as few genes are able to be activated.

### Evolution Continuation and Termination

Once the newly generated (or initial) members of the population are evaluated, the next round of evolution of the population can occur, with the chromosomes of the fittest circuits being more likely to be chosen for reproduction. This process is continued until a chromosome with 100% fitness is attained, or until a predetermined number of generations has been completed, or until there has been no increase in the maximum fitness achieved in the population for a predetermined number of generations (i.e. stagnation has occurred).

## 5.3    Chromosome Preprocessing

### Purpose

The function of the chromosome preprocessor is to parse the chromosome, locating its genes, their coding regions, promoters and binding sites on regulatory regions. The coding regions are then decoded into a list of gene products, according to the supplied genetic code, and the binding of resources and simulated molecules (if used) to the gene's binding sites is determined, using the supplied bind code to determine their bind templates.

This extracted information can then be used by a developmental system that is driven by transcription-level gene regulation. This eliminates the need to re-parse and extract, translate and collate information from the chromosome more than once. The extracted information can then be reused at all stages of the developmental process, and by each cell.

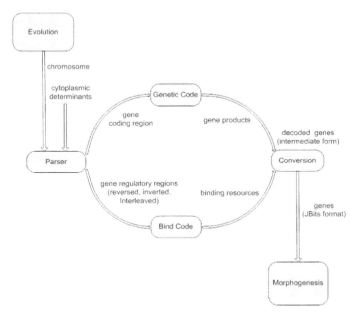

Figure 5.2: Chromosome to Morphogenetic Specification Process

## Chromosome to Morphogenesis Specification Pipeline

To go from a variable length chromosome, containing several genes separated by 'junk' regions, to a growth process specification requires several stages in a pipelined process. The stages involved are outlined in Figure 5.2. Note that the resource conversion table is not shown explicitly; it is included within the conversion process.

## Chromosome Parsing and Decoding

Firstly, the chromosome preprocessor parses the chromosome to locate genes, their coding and regulatory regions. Genes are comprised of a coding region, optionally preceded by a promoter site and regulatory regions (enhancers and repressors). Enhancers and repressors are determined by their location on the gene relative to the promoter, with enhancers being located upstream of promoter, where they act to attract the polymerase enzyme for gene transcription, while repressors are located between the promoter and gene coding region, thus blocking polymerase from transcription (from a conceptual view as illustrated in Section 4.4.3).

These regions are then extracted and decoded using, respectively, a genetic code and bind code to produce an intermediate text-based representation of the FPGA resources that are to be configured, for gene products, or queried for testing binding of FPGA resources to regulatory regions. See Table C.1 and the related resource codes in Appendix C.1 for details of the genetic code used for the experiments presented in this book; the bind code is directly derived from this, as explained in Section 4.4.2.

## Chromosome Grammar

The following grammar is used to extract and decode genes. Note that gene coding regions have higher precedence than promoters, in that any promoter found within a coding region is treated as part of the coding region rather than as a promoter. Also, genes without promoters aren't used in the morphogenesis process, reflecting their low level of activation without a promoter to encourage the binding of polymerase.

The chromosome is parsed from left to right according to the following grammar to produce a list of genes with associated regulatory regions:

```
<chromosome>= <genecoding>* (<genectrl> <genecoding>+)* <genectrl>?
<genecoding> = STARTCODON <codon>* <stopcodon>
<genectrl>   = <enhancer>? PROMOTER <repressor>?
<enhancer>   = <regulator>
<repressor>  = <regulator>
<codon>      = <base> <base> <base>
<base>       = 0 | 1 | 2 | 3 (RNA equiv U | C | A | G)
<stopcodon>  = STOP_AMBER | STOP_OCHRE | STOP_OPAL
PROMOTER     = 0202                      (DNA equiv TATA)
STARTCODON   = 203                       (RNA equiv AUG)
STOP_AMBER   = 023                       (RNA equiv UAG)
STOP_OCHRE   = 022                       (RNA equiv UAA)
STOP_OPAL    = 032                       (RNA equiv UGA)
```

The gene coding region is now specified by the codons in *genecoding*, however the regulatory regions need to be further processed. The *regulators* are then parsed from right to left, with 2 of each 3 bases sampled to generate 3 lists of codons (starting at offsets 0, 1 and 2) which can then be decoded to locate bind sites:

```
<regulator>  = <bindcoding> <base>? <base>?
             & <base> <bindcoding> <base>?
             & <base> <base> <bindcoding>
<bindcoding> = <bcodon>+
<bcodon>     = <cbase><cbase><ignorebase><cbase>
```

```
<cbase>      = <base>(base is complemented)
<ignorebase> = <base>(base is discarded)
```

The 3 extracted *bindcoding* lists then have the *ignore* token in each *bcodon* discarded, and the *cbase* tokens are replaced with their complement ($0 \leftrightarrow 2$, $1 \leftrightarrow 3$) to give a list of *codons*, which is renamed to *bindcodons*. This is then parsed, to locate bind sites, as follows:

```
<bindcodons>  = <codon>* (<bindsite> <codon>*)*
<bindsite>    = <bstartcodon> <codon>* <bstopcodon>
<bstartcodon> = 000 | 001 | 020 | 021
<bstopcodon>  = <stopcodon> | 222 | 223 | 232
```

## Gene Conversion to JBits Format

Once the chromosome has been decoded, the resulting intermediate form of gene products and binding resources is then converted to another format that specifies the JBits constant names for FPGA gate-level logic resources. This text-based genetic format can then be used with a text-based JBits interface.

This conversion process is also responsible for removing underspecified entries, mapping resource types to the JBits resources available in the CLB's slice or logic element to which the cell is mapped, and optionally mapping resources or settings to a subset of those available. This is done using a resource conversion table, with different tables being able to be specified, to map resources to subsets of those available according to what is required for given evolutionary run. The details of this conversion, for the experiments presented in this book, are given in Appendix C.2.

The resulting decoded chromosome is then passed to the morphogenesis system which creates data structures to represent this gene expression model, that are then used to direct the morphogenesis process.

## Example Decoded Gene

An example of a decoded gene (formatted for readability), comprised of a gene coding region, a promoter site and an enhancer bind site, is given below. Each line contains the fields: chromosome number ":" gene number "," gene start (from start codon) ":" feature (one of gene, prom, enhb or repb) "," +/-offset from gene start "," length of region ":" details. The details field for bind sites (enhb or repb) and gene coding regions is a list of semi-colon delimited elements, with each element itself being comma-delimited and containing the offset, length and element details.

```
1: 1,60: gene,0,24: 3,9,SliceIn,G4,11;12,9,SliceIn,CE,4
1: 1,60: prom,-23,4: 0202
1: 1,60: enhb,-56,18: 0,8,LUTBitFN,F,0010;2,4,OutToSingleBus,,;
```

113

```
2,9,OutToSingleBus,0,5,1;3,4,OutToSingleBus,,;
3,9,SliceToOut,7,0;4,9,OutToSingleBus,0;
9,8,OutToSingleDir,connect_E,9,0
```

*connect_* prepended attributes in the bind regions indicate that the resource to be queried for binding is located in another CLB, and so it is a connection from another CLB that will be queried. For example a CLB input (`SliceIn`) resource would query the associated `OutToSingleBus` or `OutToSingleDir` single line. This is used for implementing inter-cellular signaling pathways. Note that cells that are within the same CLB can use shared CLB-wide resources to achieve the same results.

## Example JBits Converted Gene

The result of converting the above decoded gene to its JBits specifying equivalent is shown below. As cells were mapped to logic elements, with one LUT and two out bus lines each, there are four possible JBits resources that the same FPGA resource could be mapped to. Each of these is specified, separated by a "/". Also, as slice control lines are unused, due to the logic element cell mapping, these are mapped to LUT inputs. The *S1G1* LUT has less input lines, which has meant in this example that there is no corresponding line available, and so this setting will be ignored by slice 1 *G* LUT-based cells (denoted elsewhere in this book as *S1G-Y* for the slice number, LUT name and LUT output). The same principle applies to the *OutToSingleDir* connection that has no corresponding *S0F-X* entry for *OutMuxToSingle*. Note, however, that these mappings could be done differently, for example using wrapping (i.e. modulus the number of lines available) so that there is always a corresponding FPGA resource.

```
1,60:gene,0,24:
    3,9,jbits=S0G4.S0G4.S0G4.SINGLE_EAST4/S0F4.S0F4.S0F4.SINGLE_EAST19/
            S1G1.S1G1.S1G1.SINGLE_EAST22/S1F1.S1F1.S1F1.SINGLE_EAST16;
    12,9,jbits=S0G4.S0G4.S0G4.SINGLE_WEST11/S0F4.S0F4.S0F4.SINGLE_WEST3/
             -/S1F1.S1F1.S1F1.SINGLE_WEST14
1,60:prom,-23,4:0202
1,60:enhb,-56,18:
    2,9,jbits=OutMuxToSingle.OUT0_TO_SINGLE_SOUTH3,OutMuxToSingle.ON/
            OutMuxToSingle.OUT2_TO_SINGLE_SOUTH5,OutMuxToSingle.ON/
            OutMuxToSingle.OUT4_TO_SINGLE_EAST14,OutMuxToSingle.ON/
            OutMuxToSingle.OUT3_TO_SINGLE_NORTH9,OutMuxToSingle.ON;
    3,9,jbits=OUT1.OUT1.OUT1.OFF/OUT5.OUT5.OUT5.OFF/
            OUT6.OUT6.OUT6.OFF/OUT7.OUT7.OUT7.OFF;
    9,8,connect=jbits,
            0,-1,OutMuxToSingle.OUT1_TO_SINGLE_EAST3,OutMuxToSingle.OFF/
            0,0,-/
            0,-1,OutMuxToSingle.OUT4_TO_SINGLE_EAST14,OutMuxToSingle.OFF/
            0,-1,OutMuxToSingle.OUT3_TO_SINGLE_EAST11,OutMuxToSingle.OFF
```

# 5.4    Morphogenesis

## Morphogenesis relation to Chromosome Preprocessing

Once the chromosome has been decoded and converted to JBits format, the decoded genes, including promoter and bind sites, are passed to the morphogenesis process. This information is then used to create a gene model, to perform the morphogenesis (growth) process on the FPGA bitstream, that is driven by the states of the resources on the FPGA, and optionally also simulated transcription factors (which includes pre-placed cytoplasmic determinants and morphogens) in the cell. The use of simulated transcription factors (which are optional), referred to as TFs, will be covered later in Chapter 7.

## Morphogenesis Process

The morphogenesis process is produced by stepping through time in a discrete manner, at each time step updating all the cells, in terms of gene activation and transcription, protein generation and binding, in the system according to an ordering defined at run time.

## Organism Structure and Cell Contents

Cells are arranged as a 2-dimensional matrix, with each mapping to an individual CLB slice or logic element on the Virtex FPGA. Cells are simulated in software, with each containing a copy of the decoded chromosome in the form of a set of genes, a number of polymerase enzymes, and TFs if they are used.

## Morphogenesis Cell Update Order

Each growth step is initiated by the circuit evaluation process. This causes each cell to be updated in a predetermined order. Update order is important, as cell updates are not atomic. The scheme used in the MGEHW system is to create an update list that starts at each of the input and then output locations on the cell matrix (corresponding to circuit inputs and outputs), and then from each of these points in turn (i.e. in a similar manner to a breadth first search), spread outwards to each neighbouring cell, starting from the cell to the East and then working clockwise through each of the other 7 neighbours, as illustrated in Figure 5.3 (with "C" indicating the current, starting cell). Then once this has been done for each input and output, then iterate through each of these newly created cell spreads, repeating the same process (ignoring already visited cells), until the whole matrix has been covered.

| 6 | 7 | 8 |
|---|---|---|
| 5 | C | 1 |
| 4 | 3 | 2 |

Figure 5.3: Cell Update Order

## Cell Update Actions

The sequence of actions performed in cell updates involves firstly the updating the binding of FPGA resources and TFs (if they are used) to bind sites on gene regulators, followed by the recalculation of gene activation levels for those genes that aren't currently being transcribed. With gene activations calculated, free polymerase may randomly bind to a positively activated gene, according to its strength of activation. Then, if they are used, TFs are updated. After this has been done, then gene transcription is performed for gene's that are bound by polymerase. These actions are elaborated in further detail below (TF updates are covered in Section 7.2), and are responsible for implementing the morphogenesis process that generates circuits on the FPGA based on the interaction between FPGA configuration state and the chromosome.

## Updating Gene Regulatory Regions Bindings

Updating the binding of proteins to regulatory regions involves querying the state of the FPGA resources with signature sequences, given by the bind code, that match subsequences of the bind site. If the queried FPGA resource's setting matches that specified on the bind site, the resource is considered to be bound to that site, noting that FPGA resources can be bound to several bind sites concurrently.

## Calculating Gene Activations

A given gene's activation ($G_i$) can be calculated as

$$G_i = \frac{E_i - R_i}{E_i + R_i} \tag{5.4}$$

where $E_i$ and $R_i$ are the gene's enhancer and repressor activation metrics, as given by Equation 5.5 below. The sign in the result indicates whether the gene is repressed (negative activation level) or enhanced, while the magnitude gives an indication of the relative strength of the more strongly activated regulatory region's importance.

Each regulator, whether enhancer or repressor, has an associated metric

**116**

$(M)$, and this is calculated as

$$M = \frac{Nb}{\sqrt{\sum_{i=1}^{Nb} D(B_i)}} \tag{5.5}$$

with $Nb$ being the number of bind sites that are bound to by one or more proteins (or TFs), and $D(B_i)$ the distance between the closest edge of the bind site $(B_i)$ and the promoter.

## Free Polymerase Binding

Once all the genes' activations have been calculated, for each free RNA polymerase, if a random threshold is exceeded (to indicate that it is ready to bind) then a gene with positive activation is randomly chosen for transcription, with any activateable gene's probability of being chosen ($P_i$ given in Eq. 5.7) proportional to its strength of positive activation ($A_i$ as given in Eq. 5.6) in relation to the sum of all positive gene activations in the cell.

$$A_i = \begin{cases} 0 & G_i \leq 0 \\ G_i & G_i > 0 \end{cases} \tag{5.6}$$

$$P_i = \frac{A_i}{\sum_{j=1}^{n} A_j} \tag{5.7}$$

## Gene Transcription

For each gene that is bound to by polymerase, transcription is performed. This involves having the polymerase move along the gene coding region, transcribing one base at a time to generate codons which are concurrently translated into FPGA settings or TFs using the genetic code. Transcription of genes is done at a fixed rate, that is a number of bases per cell update (as specified by a system parameter), and as each gene product is translated it is released into the cell by configuring the associated FPGA resource or creating the required TF.

Note that translation is effectively incorporated into transcription, by the preprocessing of the genes that was done before. Also, even though polymerase binds to the promoter to initialise transcription, when transcription begins it is done from the start of the coding region.

**117**

## 5.5    Hardware Morphogenesis Interface

### Protein-Hardware Synchronisation

The EHW morphogenesis process is driven by the gate level configuration state of the FPGA. Thus when proteins, representing FPGA resources, are released in the cell during gene transcription, the FPGA's configuration bitstream must be updated. Likewise, when updating the binding of proteins to gene bind sites, the bitstream must be queried to determine if the resource's current setting matches the setting required for a match to occur. Remembering that each FPGA resource available in the underlying region of the FPGA is represented by a single unique protein, which will have several possible setting values possible of which only one may be active at any given time. For a binding match to occur requires this setting value to match that encoded on the chromosome.

### Text-based to JBits Class and Constant Conversion

All interfacing to the FPGA, for configuration or testing of circuits, is done through Xilinx's JBits API. However, all proteins generated from the chromosome and all binding signatures to which proteins are to be matched against to determine binding, are in a text-based representation of JBits classes and constants. Hence a text-based JBits interface is used, which converts a textual representation of the JBits resource (bits) and setting (value) constants to their associated JBits `int` scalar and array constants, prior to calling the appropriate JBits `set()` or `get()` function, for releasing proteins in the cell or testing for protein binding, respectively. The Java reflect package (Sun Microsystems Inc. 2003) is used to dynamically access the JBits class constant fields based on their name (a String).

### Contention Prevention and Higher-level Constructs

The JBits text interface also prevents contention from occurring on the FPGA, and provides higher-level FPGA resource constructs, such as *LUT Active Functions*, which give extra functionality to gate-level constructs. Contention avoidance is done simply by testing the multiplexor setting on the other end of the single line that is to be driven, using a contentious resource lookup table, prior to configuring its output multiplexor, and if this configuration would cause contention, then this output mux is not configured.

**118**

## 5.6    Circuit Evaluation

**Tasks Performed by Circuit Evaluation**

The circuit evaluation process is responsible for instantiating the morphogenesis process, and for initiating each growth step undertaken by morphogenesis (as shown in Figure 5.1 by the *step()* arrow). It is also responsible for initialising the FPGA board and, prior to circuit evaluation, for updating the FPGA configuration by generating the partial FPGA configuration bitstream (using JBits method `getPartial()`) that keeps track of the recent changes to FPGA configuration from the morphogenesis process (as performed in the hardware morphogenesis interface), and uploading this to the FPGA board (using JBits method `setConfiguration()`).

**Growth Duration and Genotype Fitness**

At each growth step, the current circuit is evaluated to give a fitness metric, according to the given problem. Growth may be done for a set number of iterations, or may be continued according to the growing circuit's progress in performing the given task. The scheme that has been used in the experiments presented in this book, is to grow the circuit for a minimum number of growth steps ($m$), with fitness evaluated at each step, and growth continued if the maximum phenotype fitness ($F_{max}$) for this genotype increased in the last $m/2$ growth steps, or if phenotype fitness ($F_i$) is increasing (i.e. $F_i > F_{i-1}$).

Depending on the aims of the given evolutionary run, genotype fitness may be based on the highest phenotype (circuit) fitness attained during morphogenesis, an averaged phenotype fitness, or some other scheme. In the experiments presented later in this book, the scheme outlined above is used for determining the duration of growth, and the genotype fitness is given by the maximum phenotype fitness achieved ($F_{max}$).

**Circuit Evaluation**

Circuit evaluation may involve either, or both, testing the current FPGA configuration (for connectivity, for example), or applying signals to the circuit inputs and sampling the circuit outputs to determine the circuit's function. The evaluation is, of course, problem dependant, and so not covered here.

## 5.7    Summary

This chapter presented the major components of an implementation of the morphogenetic EHW model presented in Chapter 4. This MGEHW system is made up of a collection of processes that perform layout of the FPGA, generate a population of chromosomes, and evolve this population to completion.

Chromosomes are parsed and decoded and then the extracted and decoded genes are passed to the circuit evaluator, which instantiates and iterates the morphogenesis process for as long as it seems profitable. Interactions between the genes and the FPGA result in the generation of a circuit, which is assigned a fitness metric based on its performance at a given task. The phenotype's fitness is returned to the evolutionary process, which continues through the selection of fit chromosomes for breeding, and the removal of unfit chromosomes.

The following chapter will use this MGEHW system to determine if morphogenesis is able to be applied to gate-level EHW, and furthermore, if, as Chapter 3 seems to indicate, a biologically inspired morphogenesis model is able to scale evolution to the generation of larger and more complex circuits within the constraints of a commercially available FPGA.

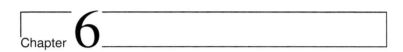

Chapter **6**

# Experiments with Evolving Circuit Structure

The aim of this chapter is to evaluate the morphogenetic model outlined in previous chapters, and to compare its performance with a traditional direct encoding approach to EHW, especially in regards scaling to increases in circuit size and complexity.

To achieve this, experiments are undertaken involving the routing of signals from one or more inputs on one edge of a CLB matrix to one or more outputs on the opposite edge, with extra constraints on connectivity to increase the difficulty of the problem, while the relationship between input and output signals is ignored.

In other words the experiments conducted in this chapter focus on evolving circuit structure, as a precursor to evolving useful functional circuits, which in any case would require that the circuits to be evolved are connected in some manner from inputs to outputs.

Due to the functional simplicity of these circuits, circuit fitness can be evaluated directly by analysing the bitstream in software, rather than requiring the bitstream to be downloaded to the FPGA at each growth step for testing. This approach minimises overhead, allowing experiments to be evaluated more rapidly, and so allowing more sets of experiments to be evaluated within the available time.

The remainder of this chapter is structured as follows. In the first section (Section 6.1), the subset of FPGA elements used in these experiments is detailed, followed in Section 6.2 by the direct encoding used for the traditional EHW approach (details of the genetic code used by the morphogenetic approach can be found in Appendix C.1, with details of the mapping to JBits resources given in Appendix C.2.1). The fitness function is then elaborated in Section 6.3, before the experiments are provided in Section 6.4. The chapter is then concluded with a summary in Section 6.5.

Table 6.1: Allocation of Resources to Logic Elements

| Single Line In | Input | Slice Out | Out Bus | Single Line Out |
|---|---|---|---|---|
| E4/N7/S18/W11 | S0G4 | S0_YQ | 0; 1 | S3; E3,N2,W4 |
| E19/N3/S9/W3 | S0F4 | S0_XQ | 2; 5 | N8,S5,S7; W16 |
| E22/N10/S9 | S1G1 | S1_YQ | 4; 6 | E14,W19; S18 |
| E16/N5/S2/W14 | S1F1 | S1_XQ | 3; 7 | N9,S10,E11; E22 |

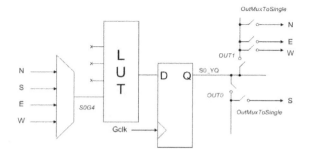

Figure 6.1: Structure of S0G-Y Slim-subset Cell

# 6.1    Resource Allocation

In these experiments, each cell is mapped to a logic element (LE), giving 4 cells to a single CLB. Each LE is then limited to a slimmed down subset of resources, with only one input used per LUT, giving 4 possible LUT functions (constant output 0 or 1, pass or invert signal).

Each LE is assigned 2 of the 8 out bus lines, each of which can connect to single lines to drive the input of a LUT in the neighbouring CLB. The set of single lines available to each logic element were chosen such that it is not possible to directly route horizontally from the West to East edges of a CLB matrix, and it is also necessary for lines to be routed through each of the 4 distinct LE types.

The allocation of resources to each LE is shown in Table 6.1, and a graphical representation of one of the four cell types is shown in Figure 6.1. Input and output cells, on the inner border of the evolvable CLB matrix are limited to $S0G-Y$ and $S1F-X$ LEs, and in the output cells, only $EAST3$ and $EAST11$ (respectively) are able to drive the circuit outputs.

## 6.2   Direct Encoding Chromosome Structure

In the direct encoding, against which the morphogenetic approach is compared, each LE is mapped, consecutively, to a group of bits on the fixed-length chromosome. Each LE configuration is specified by its single input line, LUT function, out bus mux settings, and out bus to single mux settings, in that order. These take up 4, 2, 2, 4 bits, respectively, giving each LE a 12-bit section of the chromosome encoded as *iiiiffbboooo*. This is decoded as:

iiii = input line number (0-15) Modulus number of lines + 1 (OFF)

ff = LUT FN number (0-3): output 0, pass signal, invert, output 1

bb = OUT bus Mux ($b_1$, $b_2$) setting: 0=OFF, 1=registered LE output

oooo = OUT to Single Mux ($o_i, i = 1..4$) setting: 0=OFF, 1=ON

More details can be found in Appendix C.3.1.

The ordering of the LEs on the direct encoding's chromosome is such that the LE corresponding to the first input (from outside the evolvable region) is mapped to the first section of the chromosome. From here, the remaining LEs in the evolvable region are assigned sequentially by traversing all CLBs in a CLB row from West to East, starting from the current row, then iterating through all the rows to the North before wrapping around to the South edge and iterating though the remaining rows in the same manner. Within each CLB, the order of LEs is $S0G - Y$, $S0F - X$, $S1G - Y$, $S1F - X$, and if the first LE on the chromosome is not $S0G - Y$, then the last sections of the chromosome will encode the LEs preceding this one.

## 6.3   Fitness Evaluation

Fitness is based on how much progress is made in routing a signal, possibly inverted, from the inputs to the outputs, which also requires the input and output logic elements to be connected to dedicated routing CLBs on the outside (of the evolvable region) neighbour.

The fitness metric is calculated as follows.

$$F = \frac{100}{N} \sum_{n=1}^{N} \frac{f(n)}{M} \tag{6.1}$$

where 100 is used to scale the fitness to a percentage, $N$ is the number of layers, $f(n)$ is the fitness of layer $n$ and $M$ is the width of the input and output buses (for passing signals into and out from the CLB matrix).

The fitness of layer $n$ is a function of the sum $\sigma_n$ of the connectivity value $C(l)$ of the logic elements $l$ in that layer, but with only input and output logic elements being used in the input and output layers, respectively.

$$\sigma_n = \sum_{l=1}^{L} C(l) \tag{6.2}$$

The connectivity of a logic element is assigned as follows:

| $C(l)$ | logic element connectivity |
|---|---|
| 0 | LUT input not connected |
| 1 | LUT input connected |
| 2 | LUT function passes signal (possibly inverted) |
| 3 | LUT output connected to any Out bus lines |
| 4 | Out bus line drives single connected to a LUT |

The layer fitness, $f(n)$, is calculated differently for the first, last, next to last and interior layers

$$
\begin{aligned}
f(1) &= \sigma_1 \\
f(n) &= \min(\sigma_n, 4M) * K_{p,n} \qquad 1 < n < N-1 \\
f(N-1) &= \min(\sigma_n, 4M) * p/M \\
f(N) &= \sigma_N
\end{aligned}
\tag{6.3}
$$

where $p$ is the number of sections in the layer with connected logical elements, and there are $M$ sections per layer (sections are created by splitting layers on the orthogonal axis to create $M$ equal sized groups of CLBs).
The factor $K_{p,n}$ is given as

$$
K_{p,n} = \begin{cases} 0.5 & p < M/2 \\ 0.75 & p < 3M/4 \end{cases} \qquad n >= N/2. \tag{6.4}
$$

# 6.4    Signal Routing Experiments

The aim of the experiments presented in this section, is to compare the performance of the morphogenetic and direct encoding approaches relative to increases in circuit size, as measured by the size of the CLB matrix evolved, and circuit complexity, as measured by the number of signals routed across the matrix.

All experiments in this section use the same evolutionary and morphogenetic parameters. These parameters are given in Table 6.2. A sensitivity analysis of evolutionary and morphogenetic parameters is provided in Appendix A.5.

The stagnation generations parameter, in which evolution is halted after the specified number of generations of no increases in maximum fitness achieved

Table 6.2: Signal Routing Experiment Parameters

| Parameter | Setting | Applies To |
|---|---|---|
| Maximum Generations | 10000 | MG and GA |
| Stagnation Generations | - | MG and GA |
| Population Size | 100 | MG and GA |
| Replace Rate | 10 | MG and GA |
| Crossover Rate | 80% | MG and GA |
| Mutation Rate | 2% | MG and GA |
| Inversion Rate | 5% | MG and GA |
| Base Insert/Delete Rate | 0.1% | MG only |
| Pre-evolution Generations | 25 | MG only |
| Gene Ceiling (soft) | 8 | MG only |
| Minimum Growth Steps | 30 | MG only |
| Transcription Rate | 4 | MG only |
| Polymerase to Gene Ratio | 30% | MG only |
| Polymerase Bind Probability | 20% | MG only |

by the population, was not used here; all experiments are run until a 100% solution has been found, or when 10,000 generations of evolution have been performed.

In regards the morphogenesis only evolutionary parameters, note that the base insert/delete parameter only applies to the morphogenetic approach due to its variable length chromosome; while pre-evolution, refers to the evolution of "viable" gene-containing chromosomes, discussed in Section 5.1; and the gene-ceiling parameter is used to prevent *potential* unbounded chromosome growth, that was referred to in Section 5.2.

For the remainder of this section, *MG* and *GA* are sometimes used as shorthand for the morphogenetic and direct encoding approaches to EHW, respectively. Logic element and cell are also used interchangeably, when referring to the sizes of evolved CLB regions.

## 6.4.1   Experiments in Scaling Circuit Size

In this first set of experiments, the aim is to determine the performance of both morphogenetic and direct encoding approaches to increases in circuit size, as measured by the size of the target CLB region, while the circuit function remains fixed at routing a single signal from the center of the West side of the CLB matrix to the center of the East side, while constrained to a reduced set of routing resources that disallows a direct West to East connection across the CLB matrix.

125

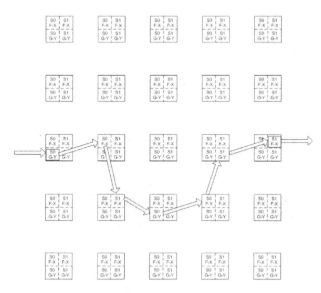

Figure 6.2: Single Signal Routing on a 5x5 CLB Matrix

### Single Signal Routing Layout

Experiments are conducted on four CLB matrix sizes, these being 5x5 (containing 100 cells), 9x9 (324 cells), 13x13 (676 cells), and 17x17 (1156 cells). This involves an increase of 11.56 times the number of logic elements from the smallest to largest evolved CLB regions. In all cases the signal input and output points remain the same to ensure that the only difference in problem difficulty lies in the scaling of CLB matrix size; the input point is LUT input single West11 into logic element $S0G - Y$, in the middle CLB row; and the output point is single $East11$ from out bus line 3 in logic element $S1F - X$, again in the middle CLB row.

Figure 6.2 illustrates, with a hand-crafted solution, the location of IO points and constraints of routing across the CLB matrix.

### Single Signal Experiment Runs

For each set of experiments 10 runs were conducted for both MG and GA approaches. Figures 6.3 and 6.4 show the mean and standard deviation of each of these sets of experiments, in terms of, respectively, generations required and chromosome length versus the number of logic elements (the number of CLBs

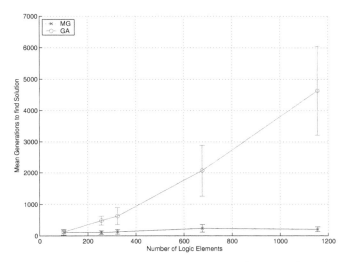

Figure 6.3: Mean and S.D. Generations Required to find Solution on Single Signal Routing

times four). For illustrative purposes, these figures also include the results from the 8x8 single signal routing experiment (which lies between the 5x5 and 9x9 experiments, with 256 cells, but has different IO points) from Section 6.4.2.

Note, however, that for the 17x17 CLB problem, one of the GA runs didn't complete in 10,000 generations, having reached a maximum fitness of around 82.35% at generation 5636. This run was not included in the mean or standard deviation of the direct encoding approach.

## Comparing Generations Required for MG vs GA

From Fig. 6.3 it appears that the number of generations required to find a solution is not dependent on the size of the evolved region for the morphogenetic approach. For the direct encoding approach, on the other hand, the number of generations required appears to be tending towards a polynomial, or exponential, function of the size of the evolved region, although more experimentation would be required to determine the exact function.

For the 5x5 experiment, the direct encoding approach took a mean of 121 generations (S.D. of 58.9) to find a 100% solution, while the morphogenetic approach took a mean of 109.4 generations (S.D. of 93.9). Thus the morphogenetic approach generally showed superior performance, in terms of generations required in this experiment set, and the performance of the MG approach

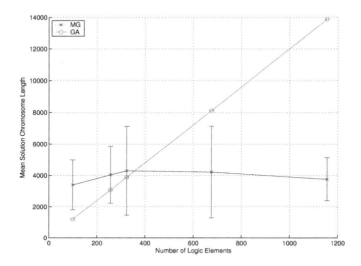

Figure 6.4: Mean and S.D. Chromosome Length for Solution on Single Signal Routing

over the GA became increasingly pronounced with increases in the CLB size.

For the 17x17 experiment, the direct encoding approach (discounting the failed run mentioned above) required on average 4635.1 generations to find the solution (S.D. of 1417.7), a 38 fold increase for the 11.56 increase in logic elements. The morphogenetic approach, however, required only 204.9 generations on average (S.D. of 75.8) to find a 100% solution, a mere 87% increase in generations required over the 5x5 problem, and furthermore the mean of the 17x17 set almost lies within 1 SD of the mean of the 5x5.

### Comparing Chromosome Length for MG vs GA

While the chromosome length of a direct encoding approach to EHW has a direct correspondence to the size of the evolvable region, it is obvious from Fig. 6.4 that this is not the case for a morphogenetic approach. While the direct encoding approach had a chromosome length that increased from 1200 bits for the 5x5 problem to 13,872 bits for the 17x17 problem, the morphogenetic approach varied from around 3500 bases for the 5x5 and 17x17 problems to around 4250 bases for the 9x9 and 13x13 problems [1]

---

[1] While strictly speaking the chromosome length for MG approaches is effectively double that of the GA due to its base-4 encoding, in reality the redundancy of the genetic code reduces this to effectively a base-2 encoding. Either way, this doesn't affect the comparison

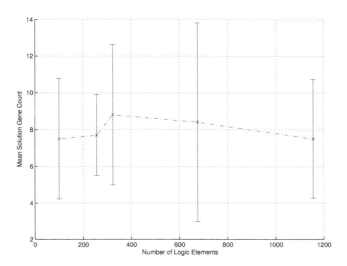

Figure 6.5: Mean and S.D. Genes in Solution on Single Signal Routing

### Comparing Gene Count against Circuit Size

Comparing the number of genes required to solve the problem, for the morphogenetic approach, also showed no correspondence with circuit size, with the 5x5 and 17x17 CLB problems both requiring on average 7.5 genes, while the 9x9 and 13x13 problems required 8.8 and 8.4 genes, respectively. In all cases the difference between experiment set means is less than the standard deviation within sets. This is illustrated in Fig. 6.5 In retrospect, the number of genes required seemed, here, to correspond closely to the chosen soft gene ceiling, which would have pushed evolution towards solutions with 8 genes, over those with more, where possible.

### Comparing Growth Steps Against Circuit Size

For the number of growth steps required to construct the solution, as illustrated by Fig. 6.6, there was no direct correspondence between circuit size and growth steps required, noting that in all cases, the minimum number of growth steps performed was 30. The 5x5 problem required on average 32.6 growth steps, the 8x8 problem (from the next section) required slightly less at 32.4, the 9x9 problem required 36 steps and the 13x13 required 38.7, but then the 17x17

---

of chromosome *growth* relative to evolved area.

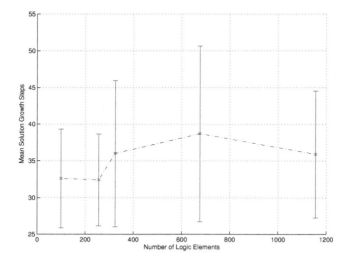

Figure 6.6: Mean and S.D. Growth Steps Required for Solution on Single Signal Routing

problem required less than both the 9x9 and 13x13 at an average of 35.9. As with the number of genes required, the difference between experiment set means was less than the standard deviation within sets.

**Morphogenetic Approach Scales to Increases in Circuit Size**

The results of these experiments clearly shows that the morphogenetic approach is able to scale to increases in circuit sizes, with little effect on generations required, chromosome length, and gene count. The number of growth steps required also appeared not to be overly dependent on circuit size, though further experiments would be required to determine this one way or the other. Also, the morphogenetic approach, on average, outperformed the direct encoding for all problems, which became increasingly obvious with increases in the size of the evolved CLB region.

**Direct Encoding Affected by Increases in Circuit Size**

Unlike the morphogenetic approach, the direct encoding approach, increasingly struggles as the circuit size is increased, and eventually at the largest problem failed to successfully complete an experiment run. It is likely that this is an indication that at even larger circuit sizes it would begin to fail more often.

## 6.4.2    Experiments in Scaling Circuit Complexity

In this set of experiments the scalability of morphogenetic and direct encoding approaches to EHW is evaluated and compared in terms of circuit complexity, by fixing the CLB matrix size to 8x8 CLBs (containing 256 cells) while increasing the number of signals routed across the matrix from 1 to 4, and hence also the number of IO points from 2 to 8.

### 8x8 Single Signal Routing Layout

In the case of single signal routing, the input point is LUT input single $West14$ into the $S1F - X$ logic element of the lower of the two middle row CLBs on the West edge of the CLB matrix, while the output point is single $East3$ from out bus line 1 in the $S0G - Y$ logic element of the upper of the two middle row CLBs on the East edge of the CLB matrix.

### 8x8 Multi-Signal Routing Layout

For the multi-signal routing experiments, inputs are placed in the center of the West edge of the CLB matrix, 2 input logic elements per CLB (LUT input single $West11$ into logic element $S0G - Y$ and $West14$ into $S1F - X$), while outputs are spread evenly across the East edge of the CLB matrix, using the $S0G - Y$ $OUT1$ to $East3$ and $S1F - X$ $OUT3$ to $East11$ logic element connections.

Here, as with the previous experiments, the aim is to route signals, possibly inverted, from inputs to outputs, but in this case it was only necessary that each input and output is fully connected. In other words, the relationship between the different inputs and outputs is not important, it is only necessary that all inputs are connected and one or more of these drive the outputs.

These experiments increase the number of IO connections by 4, while requiring evolution to learn not just how to connect horizontally across the matrix, but also how to spread vertically from the middle outwards.

### Multi Signal Experiment Runs

Again, for each set of experiments 10 runs were conducted for both MG and GA approaches. Figure 6.8 shows the mean maximum for each set at each generation, for all four sets of experiments.

### Comparing MG vs GA Performance

From Fig. 6.8 it is obvious that the morphogenetic approach outperformed the direct encoding approach for both runs. Both single and multi signal runs of the morphogenetic approach outperformed even the single signal runs of the direct encoding approach. Furthermore, there was little difference in

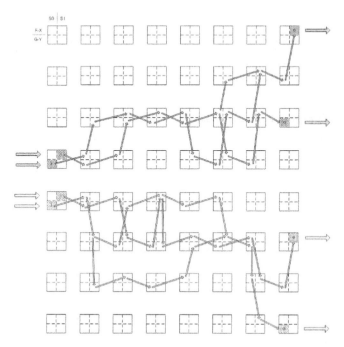

Figure 6.7: Four Signal Routing on a 8x8 CLB Matrix

performance between the single and multi signal runs for the morphogenetic approach (101.9 generations mean with S.D. of 53.8 versus 89.7 generations with S.D. of 23.1, for the single and multi signal routing sets respectively), while in comparison, the multi signal runs of the direct encoding approach took close to four times as long to solve (the single signal set had a mean of 482.5 generations and S.D. of 142.7 versus 1616.7 generations and S.D. of 598.7 for the multi signal routing set).

In terms of chromosome length, the direct encoding approach required 3072 bits for both sets, while the morphogenetic approach varied between a mean of 4043.3 bases (S.D. of 1821.9) for the single signal set and 3841.5 bases (S.D. of 1463.7) for the multi signal set.

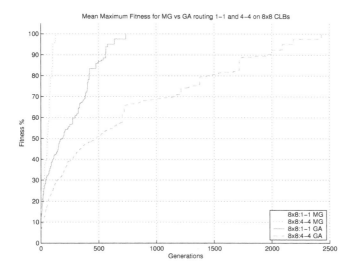

Figure 6.8: Mean Maximum Fitness for Signal Routing on 8x8 CLBs

**Gene Counts and Growth Steps**

As with the previous sets of experiments, the soft gene ceiling was set at 8 genes, and the minimum number of growth steps performed was 30. The mean gene count for solutions in both sets of experiments was 7.7 (with S.D. of 2.21 and 3.34 for the single and multi signal experiments, respectively), showing that increasing the complexity incurred no recognisable cost in terms of genes.

The growth count for solutions, however, increased slightly from a mean of 32.4 (S.D. of 6.24) for the single signal experiments to a mean of 37.9 (S.D. of 8.28).

## 6.4.3    Summary of Routing Experiment Results

Overall, the morphogenetic approach showed superior results to the direct encoding approach to the signal routing experiments. This becomes even more apparent as the experiments are scaled to larger sizes (in terms of evolved CLB region) and increased complexity (in terms of the number of signals routed), where the direct encoding approach requires an increasing number of generations to solve the given problem, while the morphogenetic approach remains relatively constant in the number of generations required, and furthermore, in-

133

Table 6.3: Summary of Signal Routing Experiment Sets

| Experiment | Max at Gen | | Chrom Length | | Growth Steps | | Genes | |
| Set | Mean | S.D. | Mean | S.D. | Mean | S.D. | Mean | S.D. |
|---|---|---|---|---|---|---|---|---|
| MG5x5 | 109.4 | 93.9 | 3393.1 | 1592.7 | 32.6 | 6.7 | 7.5 | 3.3 |
| MG9x9 | 126.3 | 72.7 | 4295.8 | 2831.7 | 36.0 | 9.9 | 8.8 | 3.8 |
| MG13x13 | 236.4 | 121.5 | 4213.5 | 2924.4 | 38.7 | 11.9 | 8.4 | 5.4 |
| MG17x17 | 204.9 | 75.8 | 3758.5 | 1366.9 | 35.9 | 8.6 | 7.5 | 3.2 |
| GA5x5 | 121.0 | 58.9 | 1200 | 0 | - | - | - | - |
| GA9x9 | 627.4 | 268.8 | 3888 | 0 | - | - | - | - |
| GA13x13 | 2070.5 | 814.4 | 8112 | 0 | - | - | - | - |
| GA17x17 | 4635.1 | 1417.7 | 13872 | 0 | - | - | - | - |
| MG8x8:1-1 | 101.9 | 53.8 | 4043.3 | 1821.9 | 32.4 | 6.2 | 7.7 | 2.2 |
| MG8x8:4-4 | 89.7 | 23.1 | 3841.5 | 1463.7 | 37.9 | 8.3 | 7.7 | 3.3 |
| GA8x8:1-1 | 482.5 | 142.7 | 3072 | 0 | - | - | - | - |
| GA8x8:4-4 | 1616.7 | 598.7 | 3072 | 0 | - | - | - | - |

curs little or no extra cost in regards chromosome length, gene count or growth steps required.

Even when the extra computational overhead is taken into account, in which the morphogenetic approach takes from 2–4 times as long as the direct encoding approach, this additional overhead is soon outweighed by the number of generations required by the direct encoding approach as soon as problem size or difficulty is increased. For a more thorough comparison of computational time required for the morphogenetic and direct encoding approaches, see Appendix A.4.

The results of all signal routing experiments are summarised in Table 6.3. Experiment sets are denoted by approach (MG or GA), CLB region (rows x columns), and for the 8x8 experiments the sets are suffixed with the number of input and output points.

## 6.5    Summary

In this chapter experiments have been conducted to compare the performance of morphogenetic and traditional direct encoding approaches to EHW to increases in circuit size and structural complexity.

The experiments involve routing signals from designated input points at one side of a matrix of CLBs to designated output locations on the opposite side. In all experiments, the routing is severely constrained to ensure that they are not trivially solvable. To route signals horizontally requires all four distinct cell types, corresponding to the different logic elements within CLBs, to be utilised, and additionally routing is forced to use vertical as well as horizontal connections to reach the other side.

Four sets of experiments were conducted involving routing a single signal across a CLB matrix, with the number of logic elements evolved rising from 100 on the 5x5 matrix, through to 1156 on the 17x17 CLB matrix. The performance of both direct encoding and morphogenetic approaches was compared, from which it was evident that with increasing size, the direct encoding approach required increasingly longer to find a solution. Furthermore, when the largest set of experiments was undertaken, the direct encoding approach began to fail. From a graph of generations required versus the number of logic elements evolved, it appears that the number of generations required appears to tend towards a polynomial, or exponential, function of the size of the evolved region.

In contrast, the morphogenetic approach was fairly constant in the number of generations required to solve the single signal routing problem, and in all cases required less generations than the direct encoding approach, on average, to find a solution.

An additional two sets of experiments were conducted on a mid-sized CLB matrix. The first involving routing a single signal across the matrix, while the second involved routing four signals from the center row across to the opposite side of the matrix to points spread across the entire edge of the matrix.

The direct encoding approach required roughly four times as many generations to solve the multi signal problem as the single signal problem (possibly corresponding to the fourfold increase in IO), while the morphogenetic approach required no extra generations for the more complex problem (the average for the four signal problem was slightly less than for the single signal problem), and in both cases solved the problem (on average) in less generations than was required by the direct encoding approach for the single signal problem.

In all experiments, the length of chromosome for the direct encoding approach was proportional to the size of the evolved CLB region, while the morphogenetic approach had no direct correspondence between CLB size or circuit complexity and chromosome length or gene count.

To conclude this chapter, the experiments conducted here clearly showed the superior performance of the morphogenetic approach both in generating circuit structures within the constraints of an FPGA, and more importantly, scaling EHW to increases in circuit size and structural complexity.

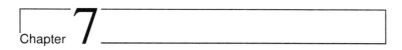

Chapter **7**

# Simulated Secondary Developmental Mechanisms

This chapter begins an investigation into whether adding simulated secondary EHW developmental mechanisms is able to further improve the performance of the morphogenetic approach. However, due to time limitations, only preliminary results are presented. Unfortunately, these results are inconclusive; further work is required to evaluate this approach to determine the value of the components in this approach, and what modifications are required to ensure success.

**Rationale For Simulated Secondary Developmental Mechanisms**

In Chapter 6 a simple morphogenetic design, based on the interactions between genes and the gate-level configuration on an FPGA, was shown to be successful at scaling EHW to larger circuits (in terms of the size of the evolved CLB matrix). This approach limits the developmental mechanisms available to those that can be implemented solely through interactions between genes and "proteins", representing FPGA gate level configuration. Furthermore, inter-cellular interactions are limited to those between neighbouring cells connected by routing resources.

While this proved to be effective, it is possible that superior performance may be gained, especially for larger CLB matrix sizes, by providing location specific information, as seen for example during pattern formation in biological organisms. In this chapter specialised simulated transcription factors that don't have any correspondence with the underlying FPGA configuration are introduced, largely, for emulating the location-specific functionality that occurs in some developmental processes.

**Chapter Outline**

First, in Section 7.1 the useful developmental mechanisms from biology that require simulated molecules are covered. Then in Section 7.2 the design and implementation of these simulated molecules is detailed. This is followed, in Section 7.3, by three sets of signal routing experiments in which these are used, and the results of these experiments are compared against the performance of the morphogenetic EHW approach which relies solely on FPGA configuration state to drive morphogenesis. Section 7.4 provides a discussion of issues with this approach based on these results, before the chapter is concluded with a summary.

# 7.1    Useful Biological Mechanisms Requiring Simulated Molecules

**Relevant Biological Processes**

There are several biological mechanisms from pattern formation and cellular differentiation, introduced in Section 4.3.2, that are relevant to morphogenetic EHW with a fixed cell-hardware mapping. These are cell-specific differentiation, axis specification and chemical gradients. The functionality provided by these is the ability to influence the patterns of gene expression, and hence development, of cells according to their location within the developing organism.

**Cell-specific Differentiation**

Generally speaking, cells all contain the same genetic information, however, their specialised structures and functionality differ according to the proteins present within the cell, and this is determined by which genes are expressed. Differentiation is the process by which different patterns of gene expression are activated and maintained to generate cells of different types.

Which genes, and hence proteins, are expressed differs between cells according to what cytoplasmic determinants are inherited at cell division, and what extracellular signals are received. Cytoplasmic determinants are molecules that bind to the regulatory regions of genes (i.e. transcription factors) and help to determine a cell's developmental fate, that is, the pattern of gene expression that causes the cell to differentiate in a particular manner.

Although cell division is not applicable to the developmental model presented in this book, the use of pre-placed simulated cytoplasmic determinants to differentiate cell types may be useful. This emulates the inheritance of locally distributed cytoplasmic determinants at cell division, which during the early stages of development helps to form regions of cells differentiating into different developmental pathways.

The most obvious use for this is to differentiate cells that correspond to the circuit input or output points. It may also be used to differentiate between the two or four distinct cell types which correspond to the two slices or four logic elements within a CLB to which cells may be mapped. Which of these a cell is assigned to is made invisible to cells through the conversion phase of chromosome decoding (see Section 5.3), however, these cells actually have slightly different resources available, and it may be advantageous to make this visible to the cell.

## Axis Specification

In the biological process of pattern formation, one of the first tasks undertaken in the embryo is to determine the principal body axis, and polarisation. This may be determined by the asymmetric distribution of maternal gene products in the egg, or require a physical cue from the environment.

For performing morphogenesis in EHW, axis specification and polarisation may also be important, and in can be provided by simulated axis specific cytoplasmic determinant molecules preplaced at run-time.

## Morphogen Gradients

Morphogens are the primary biological mechanism for determining location along an embryonic axis, and may also be used for the generation of patterns. In the developmental model presented in this book, simulated morphogens, acting as chemical gradients, may also be used for these purposes.

This can be implemented in a morphogenetic EHW system by allowing a graded response from cells according to morphogen concentration. An example of where this may be applied is to give information as to a cell's position relative to the input and output cells on the CLB matrix.

## Non-coding mRNA

Recently, it has been found that non-coding messenger RNA, which exists in higher organisms, is able to encode higher-level architectural plans (Mattick 2001). It may also be advantageous to include such a mechanism for morphogenetic EHW, so that more complex patterns of gene expression may be evolved, which don't require the manipulation of hardware resources to represent intermediate stages in the cell's state. This is analogous to the use of local variables within functions or subroutines in structured programming, as compared with being limited to only a finite number of global variables.

This can be emulated using simulated transcription factor molecules that don't form proteins (i.e. don't map to FPGA configuration settings) that are local to the cell. These molecules are able to bind to matching regulatory regions of the chromosome and hence affect gene expression. To ensure that they don't dominate gene expression over FPGA-based proteins, they are given a limited life span.

## 7.2    Details of The Implementation of Simulated Molecules

### Three Kinds of Transcription Factors

From the previous section, three kinds of simulated molecules were identified for incorporation into the morphogenetic EHW system. These were cytoplasmic determinants (axis-specific and cell-type specifiers) for cell differentiation, morphogens (diffusing chemical gradients) for determining location along an axis and generating multi-cellular patterns (of gene activation), and non-coding messenger RNA for aiding in gene regulation. These can all be implemented with simulated transcription factors (TFs), with local (to cell) TFs corresponding to non-coding messenger RNA, cytoplasm type corresponding to cytoplasmic determinants, and morphogen types.

### TF Structure and Function

Unlike "proteins" which correspond to FPGA hardware resources, TFs have no correspondence with the underlying FPGA configuration. All TFs are simply a signature sequence that is able to bind to complementary sequences of DNA on the regulatory bind sites of genes. Also, TFs differ from proteins in that there may be multiple instances of the same type within the same cell; however, they also have a limited lifespan (except for cytoplasmic determinants). Note that strictly speaking all gene products in the MGEHW system, whether FPGA resources or not, are TFs as they are all able to bind to the regulatory regions of a gene for affecting its activation.

### TF Binding

Binding of TFs to a gene's bind site only occurs if the complement of the complete TF signature sequence matches one of the reversed, folded and interleaved sequences on the bind site (see 5.3 for more details on bind sites), and no other TFs are already bound there. In the case of morphogens, the signature has a binary distance from source prepended to provide gradient information, while for cytoplasmic determinants, an optional distance from source, in the two axis, may be prepended to supply axis asymmetry information.

### Morphogen Gradient

For morphogens, the distance from source is a randomly assigned binary sequence of length $N$, where $N$ indicates the Manhattan distance from the morphogen source (by default the distance is measured in CLBs, however there are row and column diffusion parameters which indicate how many CLBs to traverse before increasing the distance metric). This sequence provides a diffusion property (i.e. chemical gradient), as when it is prepended to the TFs bind sequence the probability of it being able to bind to the adjacent region of the chromosome decreases (halves) as $N$ increases (by one). When testing for binding, the binary distance from source component of the morphogen is converted to base 4. The resulting distance sequence is then pre-pended to the TF's binding sequence, and this sequence is then matched against the regulatory region sequence's complement. In the case of binary sequences with an odd number of digits, the leading digit is treated as odd or even, rather than as a specific number, and matched accordingly (i.e. if either matches then it is able to bind).

### Morphogen Spread

When morphogens are encoded in a gene coding region, or pre-placed, then a spread from source is provided in the form of four sequential 2 digit binary strings, corresponding to the cardinal directions (in the order of S,N,W,E). Each of these uses a Gray coding ($00 = 0$, $01 = 1$, $11 = 2$, $10 = 3$) to indicate the extent of spread in that direction. The actual spread is done by following each axis out to the extent indicated by the directional spread, and then on the cells of the quadrants between each axis, the distance metric is filled up to the lowest distance specified by the two. The morphogen's distance from source is then random generated to create a binary number with as many digits as the Manhattan distance from the source.

### Morphogen Propogation

On creation morphogens are stored within all the cells to which they will propagate, along with a time to release (determined according to a user defined morphogen delay parameter) which is decremented at each growth step. When the time to release reaches zero, the morphogen is released in the cell. This simplifies and reduces the run time overhead for spreading morphogens across cells.

### Cytoplasmic Determinants

Cytoplasmic determinants, unlike morphogens and local TFs, are only created through preplacement before morphogenesis commences and so have an "eternal" life span (i.e. they last the duration of the morphogenesis run). Unlike morphogens, cytoplasmic determinants are generated in a rectangular spread (from input and output ends of the CLB matrix currently), with axis asymmetry encoded in the distance from source field of the generated TFs. The asymmetry is encoded by concatenating the same number of '1's as the Y axis (rows) distance from source with the same number of '0's as the X axis (cols) from source. This provides axis specific gradients across the cellular matrix.

Another type of cytoplasmic determinant is also provided, which lacks the distance from source. This used to differentiate between the two or four distinct cell types corresponding to the slice or logic element the cells are mapped to. These may have a distinct cytoplasmic determinant bind sequence allocated, and each cell given the corresponding cytoplasmic determinant. Cells corresponding to circuit input and output locations may also be allocated specific cytoplasmic determinants.

### Incorporating TFs into Cell Updates

To incorporate these simulated TFs into the morphogenesis model, the order of updates within a cell (as performed by all cells at each growth iteration) is updated to be as follows:

1. Update binding of FPGA resources and TFs to gene regulators.

2. Update activation levels of promoters.

3. Randomly bind some free RNA polymerase to promoter.

4. Age local TFs and morphogens.

5. Remove old TFs and morphogens with expired time to live.

6. Update morphogens' time to release.

7. Release morphogens with expired time to release.

8. Transcribe gene coding region for each active RNA polymerase.

### Integration of TFs with Chromosome Preprocessor

During chromosome preprocessing, at the same time as the list of binding FPGA resources is determined for each bind site, the bind sites are also examined for possible binding locations for local and morphogen TFs that are encoded in genes, and cytoplasmic determinants supplied from the setup phase.

This requires that all TFs, that will occur during the developmental process are known a priori, including morphogens at all possible concentrations (distances from source), to prevent an exhaustive elaboration of all possible TFs that can bind somewhere. This is required as TFs are simply variable length base-4 sequences, meaning that for each bind site there would be a significant number of possible TFs that could bind, of which few would exist during any one developmental run.

## 7.3    Experiments

In this section, experiments are conducted to determine whether the use of simulated transcription factors enhances the performance of the morphogenetic system. These experiments are based on those presented in Chapter 6. For the remainder of this section the morphogenetic approach without simulated TFs is denoted as *MG*, while the approach with simulated TFs is denoted as *TF*.

In this set of experiments, the morphogenesis system was run with support for simulated molecules (TFs) on three of the signal routing problems previously conducted without TFs (see Chapter 6). The problems chosen were the smallest (1 input to 1 output on a 5x5 CLB matrix), the problem with the most IO (4 inputs to 4 outputs on a 8x8 CLB matrix), and the next-to largest (1 input to 1 output on a 13x13 CLB matrix). For each set of experiments 10 runs were conducted, and then compared with that of the MG runs.

### TF Parameters

These experiments use the same morphogenetic and evolutionary parameters as those in which morphogenesis relied solely on FPGA configuration state, however TFs have extra parameters to determine their behaviour. These parameters are given below, and were chosen as the better of two sets of parameters tested in the preliminary experiments presented in Appendix A.3.

- Morphogen propagation delay = 1.

- Morphogen TF initial time to live = 10.

- Local TF initial time to live = 10.

- Unbound TF age rate = 5.

- Bound TF age rate = 2.

According to these parameters, morphogens will spread at a rate of 1 CLB per growth step, and a TF will live for another 2 growth steps after being released in the cell (as a gene product or through morphogen spread) if it doesn't bind to any regulatory region, or another 5 growth steps if it does bind.

### Cytoplasm TF Placement

Cytoplasmic determinant and morphogen (referred to as cytoplasms hereafter) placements were done as follows, noting that bind sequences were arbitrarily chosen based on combinations of start and stop codons. Firstly, each input and output cell was assigned a cytoplasm with no distance from source (i.e. local) and bind sequence of 022 for input cells and 203 for output cells. Then each of the 4 different cell types, according to which LE in a CLB they map to, have a cell differentiating cytoplasm with a distinct bind sequence placed in each of the cells. The allocation of these was:

**S0G-Y** 022023

**S0F-X** 023022

**S1G-Y** 203032

**S1F-X** 023032

These were also defined as being at source. Following this axis specifying, input and output layer cytoplasmic gradients were specified. These used bind sequences of 000 for input layers and 001 for outputs (these are arbitrary and are TF start codons). Distance from source on each axis is measured in CLBs from the input/output CLB. A cytoplasm spread is created that traverses the entire axis between inputs and outputs, but only covers the entire orthogonal axis on the output end (to encourage spread where it is needed). For the input half of the CLB martix, an orthogonal spread of only one quarter is created, while for the output half the spread is given as $l/n$ (where $l$ is the orthogonal length of axis, and $n$ is the number of outputs) rounded to the closest integer. The gradients are generated starting at each of the input and output cells, spreading out into the CLB matrix.

To demonstrate this, the following is a graphical representation, produced by the cytoplasm gradient production script for the 5x5 set of experiments (with 1 input and 1 output), showing the placement of cytoplasms and their distance from source on each axis. 'X' indicates that the cytoplasm is at its source, '.' represents that no placement occurred, while $R$-$C$ indicates the row ($R$) and column ($C$) distance from source. The first is generated for the input layers, and the second, for the output layers.

### Experiment Runs

A comparison of the performance of the morphogenetic and morphogenesis plus TFs approaches on the 5x5 single signal routing problem is shown in Fig. 7.3. This shows the mean, over all 10 runs, of the maximum fitness of the population at each generation.

```
      --------------------------------
  9 |  .   .   .   .   .   .   .  .   |
  8 |  .   .   .   .   .   .   .  .   |
  7 |1-01-01-11-11-21-2 .   .   .  .  |
  6 |1-01-01-11-11-21-2 .   .   .  .  |
  5 | X   X 0-10-10-20-2 .   .   .  . |
  4 | X   X 0-10-10-20-2 .   .   .  . |
  3 |1-01-01-11-11-21-2 .   .   .  .  |
  2 |1-01-01-11-11-21-2 .   .   .  .  |
  1 |  .   .   .   .   .   .   .  .   |
  0 |  .   .   .   .   .   .   .  .   |
      --------------------------------
      0  1  2  3  4  5  6  7  8  9
```

Figure 7.1: Cytoplasm Placement for Input layers


```
      --------------------------------
  9 |  .   .   .   . 2-22-22-12-12-02-0|
  8 |  .   .   .   . 2-22-22-12-12-02-0|
  7 |  .   .   .   . 1-21-21-11-11-01-0|
  6 |  .   .   .   . 1-21-21-11-11-01-0|
  5 |  .   .   .   . 0-20-20-10-1 X  X |
  4 |  .   .   .   . 0-20-20-10-1 X  X |
  3 |  .   .   .   . 1-21-21-11-11-01-0|
  2 |  .   .   .   . 1-21-21-11-11-01-0|
  1 |  .   .   .   . 2-22-22-12-12-02-0|
  0 |  .   .   .   . 2-22-22-12-12-02-0|
      --------------------------------
      0  1  2  3  4  5  6  7  8  9
```

Figure 7.2: Cytoplasm Placement for Output layers


Fig. 7.4 shows the mean, over 10 runs, of the maximum fitness of the population at each generation for both TF and MG approaches at routing 4 signals across an 8x8 CLB matrix.

Fig. 7.5, again shows the mean over 10 runs of the maximum fitness of the population at each generation for both TF and MG approaches, this time for routing a single signal across a 13x13 CLB matrix.

**145**

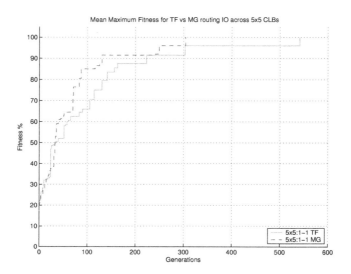

Figure 7.3: Mean Maximum Fitness for TF vs MG Approaches on Signal Routing

Table 7.1: Comparison of Morphogenetic Runs with and without Simulated TFs

| Experiment | Max at Gen | | Growth Steps | | Genes | |
| Set | Mean | S.D. | Mean | S.D. | Mean | S.D. |
|---|---|---|---|---|---|---|
| TF5x5 | 165.7 | 157.9 | 36.9 | 11.7 | 8.9 | 4.3 |
| MG5x5 | 109.4 | 93.9 | 32.6 | 6.7 | 7.5 | 3.3 |
| TF8x8:4-4 | 104.4 | 51.3 | 39.7 | 12.8 | 6.7 | 2.8 |
| MG8x8:4-4 | 89.7 | 23.1 | 37.9 | 8.3 | 7.7 | 3.3 |
| TF13x13 | 335.5 | 166.7 | 35.3 | 11.4 | 9.0 | 4.1 |
| MG13x13 | 236.4 | 121.5 | 38.7 | 11.9 | 8.4 | 5.4 |

## MG vs TF Signal Routing Result Summary

The results of the three sets of TF experiments is summarised in Table 7.1 along with the results of the morphogenesis without TFs for comparison.

From the results of these experiments it appears that adding simulated TFs adds no benefit, and actually decreases performance to a small (in relation to the direct encoding GA's performance), but noticeable degree.

Furthermore, a comparison of computational cost between the morpho-

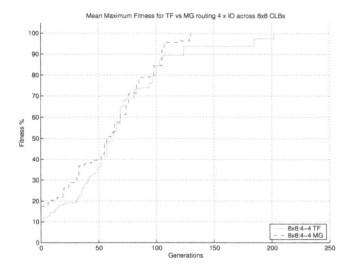

Figure 7.4: Mean Maximum Fitness for TF vs MG Approaches on Multi Signal Routing

genetic and TF approaches, presented in Table A.8 of Appendix A.4, showed that TFs required approximately 25-28% more computational time per generation than the morphogenetic approach. Perhaps, when better parameters are chosen for TFs, then an improvement in the number of generations taken to solve a given problem will offset the higher per-generation cost. This remains an open question that will require further experimentation to resolve.

## 7.4   Discussion

Overall, adding simulated transcription factors to the morphogenetic EHW approach appears to diminish performance in terms of ability to solve harder problems, and adds extra computational overhead (typically taking an extra 25% of the time required to complete the same number of generations).

### Arbitrary Bind Sequences

There were a few main areas of concern in the use of TFs. The first of these is that the cytoplasmic determinant bind sequences were chosen arbitrarily. however, at this time there is no clear manner in which bind sequences should be chosen to ensure that evolution can successfully utilise them. Specific cyto-

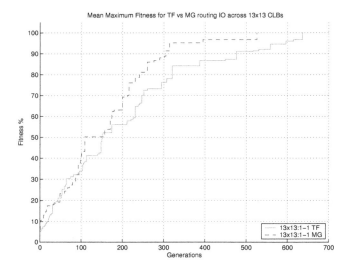

Figure 7.5: Mean Maximum Fitness for TF vs MG Approaches on Signal Routing

plasmic determinants were used at each input and output cell, and it is entirely possible that one of these interfered with the connecting of one of these cells to its required output line.

### Cost of Increased Complexity

Alternatively, or in addition to the first concern, it may simply be that adding another level of gene regulation adds an overhead in terms of the initial evolutionary learning, but which may, with increasing complexity and longer evolution, be able to outperform the simpler model.

### Placement and Spread

There is also the related issue of placement and spread of cytoplasmic determinants and morphogens, and whether they even add anything on, what is, relative to biological organisms, a fairly small region; and following from this, it is unknown if the manner in which morphogen strength is implemented, as a probabilistically decreasing likelihood of binding through the concatenation of the distance from source string, provides the desired functionality.

**Suitability of Exact Template Matching**

This brings forth another issue, which is the suitability of exact template matching for binding to regulatory sites. It may be that partial, probabilistic, matching would be better, though it would change the entire approach with morphogens. However, on the defense of this approach, it seems to work well for FPGA resources, with bind sequences determined by a redundant bind code related to the genetic code.

**Further Experiments Required For Choosing Parameters**

Finally, more effective life spans, and rates of spread for morphogens, obviously could be ascertained through further experimentation.

**Avoid Arbitrariness**

These issues would require considerable further research to resolve satisfactorily. However, at this point it appears that the results here back the premise that arbitrary mechanisms and structures add nothing, and in all likelihood, increase the likelihood of failure. This supports the morphogenetic EHW approach outlined in earlier chapters, whereby arbitrariness is avoided by coupling the morphogenesis process directly to the gate-level FPGA configuration state.

# 7.5   Summary

In this chapter, the morphogenetic model was enhanced through the addition of simulated molecules, for providing some of the location sensitive mechanisms from biological pattern formation and cellular differentiation processes that can be adapted to a fixed cell developmental process. Specifically, the additional mechanisms provided were cell-specific differentiation, from cellular differentiation, and axis specification and chemical gradients from pattern formation. Another addition was non-coding mRNA, which has been found to add an extra layer of gene regulation in higher-level organisms.

These are all implemented through simulated transcription factors (TFs), which have no correspondence with underlying hardware, but are able to directly influence the expression of genes and hence cellular development. Morphogen type TFs provide gradient information in a 2-dimensional spread around a given location. Cytoplasmic type TFs are pre-placed at run-time and provide a gradient along each axis from their source (typically cells corresponding to circuit IO). Cytoplasmic type TFs are also used for specifying cell types, such as IO cells, and for differentiating between the four different cell types that correspond to the four different logic elements within a CLB, and as such are also pre-placed at run-time. Local TFs implement non-coding mRNA.

The performance of the morphogenetic EHW system with and without TFs was compared on several signal routing experiments, the 5x5 CLBs with single input and output, 8x8 CLBs with four inputs and four outputs, and 13x13 CLBs with a single input and output. Results showed a slight deterioration in performance in all cases, and also incurred extra computational overhead.

There were several factors identified that may have contributed to this lack of performance. Primarily, the arbitrariness of cytoplasmic determinant bind sequences and placement, the implementation of morphogen gradients through the prepending of distance to generate probabilistic binding, and the evolutionary cost of an additional layer of genetic regulation. Each of these require a great deal of testing, in isolation, to determine what their actual effects are. Furthermore, more effective TF parameters need to be found through further experimentation.

These, and related, additions to the morphogenetic EHW model are an area which could possibly provide great performance improvements in the future. However, this lies beyond the scope of this book.

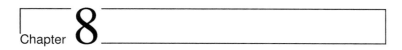

Chapter **8**

# Experiments with Evolving Circuit Functionality

While previous experiments concentrated on evolving circuits with fitness primarily based on circuit structure, using a severely cut down set of gate-level resources, in this chapter the aim is to investigate the evolution of circuit structure *and* functionality on a more complete set of resources. Concurrently, the effectiveness of various LUT encodings in aiding evolution to generate functional circuits is evaluated.

A one bit full adder was chosen as the target circuit function, being complex enough when instantiated within the constraints of an FPGA to be able to identify issues with evolving functional circuits on modern FPGAs, while simple enough to specify completely and test (reasonably) rapidly (each circuit, not genotype, evaluation taking approximately 1.6 seconds).

The LUT encodings used in the experiments are covered first in Section 8.1 (details of the genetic code used by the morphogenetic approach are given in Appendix C.1, the conversion to the subset of JBits resources is given in Appendix C.2.2, and the direct encoding is given in Appendix C.3.2), followed by the experimental setup details in Section 8.2, then circuit fitness evaluation is covered in Section 8.3, followed by the results of experiment runs in Section 8.4. As none of these experiments were able to generate a complete full adder circuit, Section 8.5 provides a hand crafted solution that demonstrates that it is possible to generate an adder within the constraints of the experimental setup. Further experiments are then undertaken in Section 8.6 to isolate and identify the problems that occurred in the initial experiments. Section 8.7 presents related work on evolving adders on a Virtex that has been reported in the literature. The chapter is summarised and concluded in Section 8.8.

## 8.1   LUT Encodings

**Bit Functions**

The native encoding of the 4-input 1-output function generators (LUTs) on
a Virtex is a 16-bit ($2^4$) inverted least significant bit first (LSB) truth table.
While this encoding, which is referred to here as *LUT Bit functions*, provides a
great deal of flexibility in expressing Boolean functions within the constraints
of an FPGA, there may be other more evolution-friendly encodings.

**Active Functions**

One such encoding is *LUT Active functions*, which provides basic Boolean
functions ($AND$, $OR$, $XOR$, $NAND$, $NOR$, $XNOR$, $Majority$ with tie
break to 0, and $Majority$ with tie break to 1) all with optional inversion of
selected lines. The designated function is applied only to currently *active* input
lines, that being inputs that are driven by active CLB outputs. For the line
to be considered active, it must be connected to a changing signal, originating
either in the circuit inputs (and generally having undergone various function
transformations), or in a feedback loop, which may generate an oscillating
signal (for example a clock divider).

This is based on the idea that a cell's functional output should be based
solely on active input signals. In other words, unconnected inputs and undriven
input lines (that are held high) should be eliminated from LUT functions,
where they may dominate the LUT's function. An example of this is a simple
OR function of all LUT inputs, which if any input line is disconnected results
in the LUT's output always being a 1, no matter what signals are received on
the other lines.

**Incremental Functions**

Another related encoding is *LUT Incremental functions*. These provide the
same set of basic Boolean functions as Active functions, again with optional
inversion of selected lines, however, unlike LUT Active functions, these are
applied to both active and inactive inputs. The inputs to which the function
is applied are chosen by a line template, which specifies whether the line is
ignored, used directly, or inverted. Furthermore, LUT Incremental functions
allow a complex function to be built up by successive applications of these
simple Boolean operations on the specified combination of inputs.

This works by combining the truth table of a simple Boolean function with
the existing function encoded in the LUT to produce a more complex LUT
function. The manner of combining the new function with the old can be
one of $SET$, $AND$, $OR$, $XOR$. The first simply sets the new LUT function
to be the simple Boolean function provided, while the others do a bitwise

$AND/OR/XOR$ of the new function's truth table and the existing LUT's uninverted truth table to produce the updated LUT truth table values. LUT incremental functions can also be applied to LUT's whose function was produced by a direct setting of the LUT bit values (by LUT Bit functions).

An example of a LUT Incremental function, is AND,OR,110- which will generate an OR function on LUT input lines 4, 3, and inverted line 2, and then $AND$ this function's truth table with the (uninverted) current LUT truth table, and store the result (inverted for storage) in the LUT.

## 8.2    Experimental Setup

### Circuit Function Definition

A one bit full adder has 3 inputs $(x, y, cin)$ and 2 outputs $(sum, cout)$, with the relation between these being defined by Boolean Equations 8.1 and 8.2.

$$
\begin{aligned}
sum &= x \oplus y \oplus cin & (8.1) \\
cout &= x \cdot y + x \cdot cin + y \cdot cin & \\
&= x \cdot y + (x + y) \cdot cin & (8.2)
\end{aligned}
$$

### Layout

The target for the adder circuit is a 2x2 CLB matrix in the center of the FPGA. This size was chosen as it contains sufficient logic and routing to provide evolution with freedom to explore a wide range of possible solutions to this problem. Although an adder could be easily defined using only two 4 input function generators (LUTs) in a single slice of a CLB, most functional circuits would require a combination of LUTs with routing lines connecting them. Thus, this approach allows the results from these experiments to be more readily generalised to other functional circuits.

The evolvable region is then setup so that the $x$ and $y$ signals are provided to single lines that can be fed to the LUT inputs in the South West CLB, while the $cin$ signals are fed towards the North West CLB's LUT inputs, and the $sum$ and $cout$ output signals are sampled from one logic element each on the two East CLBs (via OUT6 to Single East18 on the South East CLB, and OUT3 to Single East11 on the North East CLB). This layout is shown in Figure 8.1. Notice that several lines are available to the circuit inputs, while the outputs require signals to be routed to specific lines, the actual single lines connecting the IO into and out of the evolvable region are given in Table 8.1. Signals are routed directly from IO blocks on the perimeter of the FPGA to the CLBs adjacent to the evolvable region. This approach is an artefact of

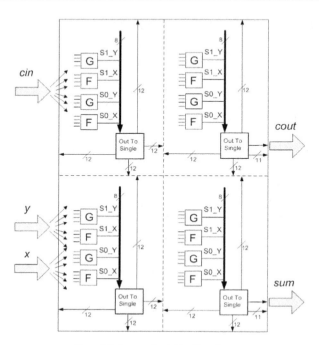

Figure 8.1: Layout for Adder Experiments

the original design aims, which were to allow the evolution of asynchronous circuits for robot control.

Though it is not shown in Fig. 8.1, all the registers in the evolvable region have their Clk line tied to a global clock (this is shown in Fig. 8.2). While the registers aren't utilised by evolution in these experiments, the contents of the registers are used for testing the activity of logic elements (LEs) in the connectivity analysis of the circuit conducted by the fitness function.

**Cell-Hardware Mapping**

As was done in the previous experiments, cells again correspond to logic elements (4 per CLB), each containing a LUT-register pair, however, in this case the CLB out bus lines are shared by all LEs in a CLB. Only the unregistered slice outputs are manipulated in these experiments due to the combinatorial nature of the adder circuit, and as cells are mapped to LEs, slice control lines are not used, being mapped to LUT inputs instead, and slice Clk lines are

**154**

Table 8.1: Adder IO Connection Points

| Signal | Connecting Single | Evolvable Connection Point |
|--------|-------------------|----------------------------|
| $x$ | West2 | S0F2 S0G2 S1F3 S1G3 |
| | West3 | S0F1 S0G1 S1F4 S1G4 |
| | West5 | S0F4 S0G4 S1F1 S1G1 |
| | West6 | S0F3 S0G3 S1F2 S1G2 |
| | West8 | S0F3 S0G3 S1F2 S1G2 |
| | West11 | S0F4 S0G4 S1F1 S1G1 |
| $y$ | West14 | S0F4 S0G4 S1F1 S1G1 |
| | West15 | S0F3 S0G3 S1F2 S1G2 |
| | West17 | S0F1 S0G1 S1F4 S1G4 |
| | West18 | S0F1 S0G1 S1F4 S1G4 |
| | West20 | S0F2 S0G2 S1F3 S1G3 |
| | West23 | S0F3 S0G3 S1F2 S1G2 |
| $cin$ | West2 | S0F2 S0G2 S1F3 S1G3 |
| | West3 | S0F1 S0G1 S1F4 S1G4 |
| | West5 | S0F4 S0G4 S1F1 S1G1 |
| | West6 | S0F3 S0G3 S1F2 S1G2 |
| | West8 | S0F3 S0G3 S1F2 S1G2 |
| | West11 | S0F4 S0G4 S1F1 S1G1 |
| $sum$ | East3 | OUT1 |
| $cout$ | East11 | OUT3 |

unused. All directly connectable single lines, and dedicated horizontally connecting out lines (OUT0 and OUT1 connecting to the east neighbour, and OUT6 and OUT7 connecting to the west neighbour) are used. Recurrent connections within a CLB (that is directly connecting LUT outputs to LUT inputs) are disallowed, however.

The available resources to a logic element are shown in Figure 8.2, with layout resources also shown and out bus mux inputs from the other LEs in the CLB. Also, the connections from the *Out To Single* multiplexor (which is responsible for driving the logic element outputs to neighbouring cells) are aggregated for the CLB. Although this only shows one of the four LE cell-types, the others have an equivalent set of resources available.

Note that the set of lines available to each LE in a CLB is the same. The lines available to each LUT input are given in Table 8.2.

The lines directly connecting to neighbouring CLBs, available from each out bus line, are provided in Table 8.3, with lines that are susceptible to contention marked with an asterisk. While 16 of the 48 single lines that are being used are able to be driven by the outputs of two neighbouring CLBs, potentially causing contention which may damage the FPGA, the contention

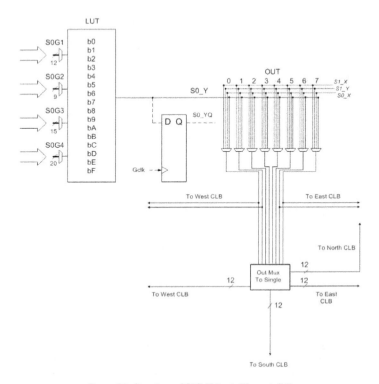

Figure 8.2: Structure of S0G-Y Logic Element Cell

avoidance mechanism provided by the text-based interface to JBits (see Section 5.5) ensures that this never occurs.

### Circuit Evaluation Platform

For the experiments presented in this chapter the *VirtexDS* simulator that is provided with JBits 2.8 is used, with a simulated Virtex XCV50 and a speed grade of 6. A simulator was used due to the unavailability of a dedicated FPGA development platform during the development of the morphogenetic EHW system, and the need to run several experiments concurrently (due to time constraints).

Due to the use of the VirtexDS simulator, recurrent connections to a CLB input from its unregistered outputs are disallowed, as they are unresolvable.

Table 8.2: Available LUT Inputs

| LUT Input Muxes | Directly Connecting Input Lines |
|---|---|
| S0G1 S0F1 S1G4 S1F4 | OFF |
| | N15 N19 N22 |
| | S6 S12 S13 S21 |
| | W5 W17 W18 OUT_WEST1 |
| S0G2 S0F2 S1G3 S1F3 | OFF |
| | E11 E23 |
| | N12 N17 |
| | S1 |
| | W2 W20 OUT_WEST0 |
| S0G3 S0F3 S1G2 S1F2 | OFF |
| | E5 E7 E9 E10 E21 OUT_EAST7 |
| | N13 |
| | S0 S14 S20 |
| | W6 W8 W15 W23 |
| S0G4 S0F4 S1G1 S1F1 | OFF |
| | E4 E16 E17 E19 E22 OUT_EAST6 |
| | N0 N1 N3 N5 N7 N10 |
| | S2 S8 S9 S18 |
| | W3 W11 W14 |

Table 8.3: Available CLB Outputs

| OUT Bus | Directly Connecting Output Lines |
|---|---|
| 0 | OUT0 (East), E2, N0*, N1*, S1*, S3, W7 |
| 1 | OUT1 (East), E3, E5*, N2, S0*, W4, W5* |
| 2 | E6, N6, N8, S5, S7, W9 |
| 3 | E8, E11*, N9, S10, W10, W11* |
| 4 | E14, N12*, N13*, S13*, S15, W19 |
| 5 | E15, E17*, N14, S12*, W16, W17* |
| 6 | OUT6 (West), E18, N18, N20, S17, S19, W21 |
| 7 | OUT7 (West) E20, E23*, N21, S22, W22, W23* |

Unfortunately, there are also some other circuit configurations that may cause the simulator to fail, in these cases the chromosome is assigned a fitness of zero.

Aside from the issues mentioned above, evolution of adder circuits in simulation or on real hardware should provide the same results due to the immunity of combinatorial circuits to timing issues (gate delays and signal ordering) in their evaluation.

## 8.3   Fitness Evaluation

Fitness is based on how much progress is made in connecting from the inputs to the outputs, *and* on how closely the connected outputs match that of the desired function. Both connectivity ($c$) and functional adder fitness ($a$) components are given as a percentage (i.e. in the range 0 to 100 inclusive). The circuit fitness ($f$) is given by adding these together and then scaling the result back to a percentage (i.e. from 0 to 100), hence

$$f = (c + a)/2. \tag{8.3}$$

### Circuit Connectivity Evaluation

The circuit connectivity portion of fitness is produced in one of two ways, depending on which of the LUT encoding methods is used. If LUT Active functions are used, then a recursive connectivity test is done based on the FPGA's current configuration. Fitness is calculated in the same manner as was done for the previous signal routing experiments (see Section 6.3). Briefly, connectivity fitness is calculated by testing how many CLB layers are connected, and for each connected layer, how many elements in the layer are connected, except in input and output CLB layers where only the connectivity of the assigned input and output LEs is utilised by the fitness function [1].

For Active functions, there are 5 levels of connectivity measurable within a logic element. However, if LUT Active functions aren't used, then each logic element is only assigned a connectivity value based on whether or not there are signal changes (indicating connectivity) on the probe at the LE's registered output, when signals are applied to the circuit input bus. Thus, for LUT Bit or Incremental functions, there are only 2 levels of connectivity measurable within a logic element.

### Circuit Functionality Evaluation

The adder circuit function fitness component is evaluated only when one or more circuit outputs are connected. To evaluate the circuit's functional fitness, each of the possible input vectors is applied and the connected circuit outputs are tested to determine if the circuit response is correct, as defined by Equations 8.1 and 8.2.

---

[1] Unfortunately, in the adder experiments there are only these 2 layers, so there was less connectivity information to guide evolution than was expected. This was an unintended consequence of re-using the connectivity analyser from the signal routing experiments.

Function equivalence is measured as the Hamming distance between the circuit outputs that *change* and the defined adder outputs. Unchanging circuit outputs are assumed to be unconnected to the inputs, and for each of these 1 is added to the Hamming distance for each input vector applied. Adder function fitness is then based on the proportion of matching signals, calculated as

$$a = 100(1 - \frac{h}{l}) \qquad (8.4)$$

where $h$ is the Hamming distance (with an integer value from 0 to $l$), and $l$ is given by

$$l = O \cdot 2^I \qquad (8.5)$$

with $O$ being the width of the circuit output bus, and $2^I$ is the number of input vectors required to completely specify each circuit output's truth table given a circuit input bus width of $I$. By substituting $I = 3$ ($x$, $y$ and $cin$) and $O = 2$ (*sum* and *cout*) into Eq. 8.5, this gives the length of the circuit test vector as

$$l = 2 \cdot 2^3 = 16. \qquad (8.6)$$

## 8.4   Experimental Runs

All experiments in this section use the same evolutionary and morphogenetic parameters as used in the previous signal routing experiments (but with a different gene ceiling and stagnation parameters). These parameters are given in Table 8.4 for convenience.

### Evolution with Different LUT Encodings

Two evolutionary runs were done for each of the three LUT encodings and a directly encoded GA with LUTs natively encoded (computational expense and time limitations prevented more than this), using the evolutionary and morphogenetic parameters given in Table 8.4.

Figure 8.3 shows the mean maximum fitness over all runs, for up to 5000 generations, for each LUT encoding approach. Although these runs are not statistically significant, they can be seen as a rough indication of trends for the different approaches.

Unfortunately, no approach was able to find a 100% solution, the best achieved was 96.875% (which means only 1 of the 16 output signals was incorrect), which was achieved by both Active LUT runs at generations 670 (growth step 31 with 12 genes) and 1849 (growth step 68 with 6 genes), and the direct encoding GA approach at generations 174 and 536. The LUT Incremental function encodings reached maximums of 90.625% and 93.75% (corresponding to 3 and 2 wrong signals) at generations 2284 (growth step 37 with 3 genes)

Table 8.4: Adder Experiment Parameters

| Parameter | Setting |
|---|---|
| Population Size | 100 |
| Replace Rate | 10 |
| Crossover Rate | 80% |
| Mutation Rate | 2% |
| Inversion Rate | 5% |
| Base Insert/Delete Rate | 0.1% |
| Stagnation Generations | 2000 |
| Maximum Generations | 5000 |
| Pre-evolution Generations | 25 |
| Gene Ceiling (soft) | 12 |
| Minimum Growth Steps | 30 |
| Transcription Rate | 4 |
| Polymerase to Gene Ratio | 30% |
| Polymerase Bind Probability | 20% |

and 2973 (growth step 32 with 8 genes). The native (LUT Bit function) encodings were only able to achieve fitnesses of 93.75% and 87.5% (corresponding to 2 and 4 wrong signals) at generations 4128 (growth step 67 with 2 genes) and 1369 (growth step 33 with 13 genes), respectively.

LUT Active functions, not surprisingly, were able to find better solutions more easily than the other morphogenetic approaches. On the other hand, there seemed to be little difference between LUT Incremental functions and Bit functions, indicating that the native encoding was no harder for evolution to explore than the contrived encoding.

## Direct Encoding Performance

Surprisingly, the direct encoding approach of traditional EHW was able to find good solutions faster than the morphogenetic approach, although it too failed to find 100% solutions. However, it should be noted that, while the GA used the same evolutionary parameters as the morphogenetic approach, its mutation rate is effectively much higher, as *most* mutations will have an effect, whereas for the morphogenetic approach the majority of mutations will have no effect, and those that do, will tend to cause smaller mutations due to the nature of the genetic encoding. Crossover too would be likely to have less pronounced effects for the morphogenetic approach, due to its use of homologous crossover. The result of this is that a direct encoding approach using the same evolutionary parameters is likely to explore the solution space a lot faster than the morphogenetic approach presented in this book.

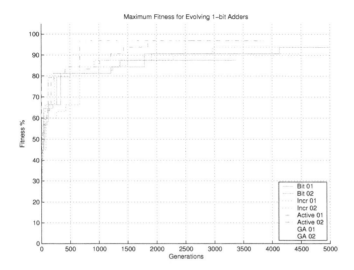

Figure 8.3: Maximum Fitness for Evolving 1-bit Adders

Furthermore, it is possible that within the constraints of the small region allocated for evolution (4 CLBs), there is little room for a morphogenesis process to provide much additional assistance to circuit generation.

## 8.5    Hand-Crafted Solution

To ensure that the problem was actually solvable within the constraints of the experiment, a handcrafted solution to the one bit full adder was generated, and tested on the same layout bitstream as used in the experiments, using Boolean Equations 8.1 and 8.2, and implemented with the following settings. It was, as expected, assigned a fitness of 100% showing that the problem should be solvable.

Note that "#" begins a commented out line, used here for an alternate LUT function implementation.

```
set: 7,11: jbits=SOG4.SOG4,SOG4.SINGLE_WEST3
set: 7,11: jbits=SOG3.SOG3,SOG3.SINGLE_WEST15
set: 7,11: jbits=SOG2.SOG2,SOG2.SINGLE_NORTH12
set: 7,11: jbits=SOG1.SOG1,SOG1.OFF
set: 7,11: LUTBitFN=LUT.SLICE0_G,1100001100111100
```

161

```
#set: 7,11: LUTIncrFN=LUT.SLICE0_G,SET,XOR,111-
#set: 7,11: LUTActiveFN=LUT.SLICE0_G,XOR,1111
set: 7,11: jbits=OUT3.OUT3,OUT3.SO_Y
set: 7,11: jbits=OutMuxToSingle.OUT3_TO_SINGLE_EAST11,OutMuxToSingle.ON

set: 7,11: jbits=S1G4.S1G4,S1G4.SINGLE_WEST5
set: 7,11: jbits=S1G3.S1G3,S1G3.SINGLE_WEST20
set: 7,11: jbits=S1G2.S1G2,S1G2.OFF
set: 7,11: jbits=S1G1.S1G1,S1G1.OFF
set: 7,11: LUTBitFN=LUT.SLICE1_G,1111111111110000
#set: 7,11: LUTIncrFN=LUT.SLICE1_G,SET,AND,11--
#set: 7,11: LUTActiveFN=LUT.SLICE1_G,AND,1111
set: 7,11: jbits=OUT1.OUT1,OUT1.S1_Y
set: 7,11: jbits=OutMuxToSingle.OUT1_TO_SINGLE_EAST3,OutMuxToSingle.ON

set: 7,11: jbits=S1F4.S1F4,S1F4.SINGLE_WEST5
set: 7,11: jbits=S1F3.S1F3,S1F3.SINGLE_WEST20
set: 7,11: jbits=S1F2.S1F2,S1F2.OFF
set: 7,11: jbits=S1F1.S1F1,S1F1.OFF
set: 7,11: LUTBitFN=LUT.SLICE1_F,1111000000000000
#set: 7,11: LUTIncrFN=LUT.SLICE1_F,SET,OR,11--
#set: 7,11: LUTActiveFN=LUT.SLICE1_F,OR,1111
set: 7,11: jbits=OUT0.OUT0,OUT0.S1_X
set: 7,11: jbits=OutMuxToSingle.OUT0_TO_SINGLE_NORTH1,OutMuxToSingle.ON

set: 7,12: jbits=S0G4.S0G4,S0G4.SINGLE_WEST11
set: 7,12: jbits=S0G3.S0G3,S0G3.OFF
set: 7,12: jbits=S0G2.S0G2,S0G2.OFF
set: 7,12: jbits=S0G1.S0G1,S0G1.OFF
set: 7,12: LUTBitFN=LUT.SLICE0_G,1111111100000000
#set: 7,12: LUTIncrFN=LUT.SLICE0_G,SET,OR,1---
#set: 7,12: LUTActiveFN=LUT.SLICE0_G,OR,1111
set: 7,12: jbits=OUT1.OUT1,OUT1.SO_Y
set: 7,12: jbits=OutMuxToSingle.OUT1_TO_SINGLE_EAST3,OutMuxToSingle.ON

set: 7,12: jbits=S0F4.S0F4,S0F4.SINGLE_WEST3
set: 7,12: jbits=S0F3.S0F3,S0F3.OFF
set: 7,12: jbits=S0F2.S0F2,S0F2.OFF
set: 7,12: jbits=S0F1.S0F1,S0F1.OFF
set: 7,12: LUTBitFN=LUT.SLICE0_F,1111111100000000
#set: 7,12: LUTIncrFN=LUT.SLICE0_F,SET,OR,1---
#set: 7,12: LUTActiveFN=LUT.SLICE0_F,OR,1111
set: 7,12: jbits=OUT0.OUT0,OUT0.SO_X
set: 7,12: jbits=OutMuxToSingle.OUT0_TO_SINGLE_NORTH1,OutMuxToSingle.ON

set: 8,11: jbits=S0G4.S0G4,S0G4.SINGLE_WEST3
```

162

```
set: 8,11: jbits=S0G3.S0G3,S0G3.OFF
set: 8,11: jbits=S0G2.S0G2,S0G2.OFF
set: 8,11: jbits=S0G1.S0G1,S0G1.OFF
set: 8,11: LUTBitFN=LUT.SLICE0_G,1111111100000000
#set: 8,11: LUTIncrFN=LUT.SLICE0_G,SET,OR,1---
#set: 8,11: LUTActiveFN=LUT.SLICE0_G,OR,1111
set: 8,11: jbits=OUT5.OUT5,OUT5.SO_Y
set: 8,11: jbits=OutMuxToSingle.OUT5_TO_SINGLE_SOUTH12,OutMuxToSingle.ON

set: 8,11: jbits=S1F4.S1F4,S1F4.SINGLE_WEST5
set: 8,11: jbits=S1F3.S1F3,S1F3.SINGLE_SOUTH1
set: 8,11: jbits=S1F2.S1F2,S1F2.OFF
set: 8,11: jbits=S1F1.S1F1,S1F1.OFF
set: 8,11: LUTBitFN=LUT.SLICE1_F,1111111111110000
#set: 8,11: LUTIncrFN=LUT.SLICE1_F,SET,AND,11--
#set: 8,11: LUTActiveFN=LUT.SLICE1_F,AND,1111
set: 8,11: jbits=OUT6.OUT6,OUT6.S1_X
set: 8,11: jbits=OutMuxToSingle.OUT6_TO_SINGLE_EAST18,OutMuxToSingle.ON

set: 8,12: jbits=S1F4.S1F4,S1F4.SINGLE_WEST18
set: 8,12: jbits=S1F3.S1F3,S1F3.SINGLE_SOUTH1
set: 8,12: jbits=S1F2.S1F2,S1F2.OFF
set: 8,12: jbits=S1F1.S1F1,S1F1.OFF
set: 8,12: LUTBitFN=LUT.SLICE1_F,1111000000000000
#set: 8,12: LUTIncrFN=LUT.SLICE1_F,SET,OR,11--
#set: 8,12: LUTActiveFN=LUT.SLICE1_F,OR,1111
set: 8,12: jbits=OUT3.OUT3,OUT3.S1_X
set: 8,12: jbits=OutMuxToSingle.OUT3_TO_SINGLE_EAST11,OutMuxToSingle.ON
```

## 8.6    Experiments Isolating the Causes of Failure

### Function Evaluation Variants

While the initial experiments failed to generate an adder function, it was found to be trivial to achieve 100% connectivity between circuit inputs and outputs. For this reason, it was first thought that the problem lay in the function evaluation portion of the fitness function, which was based on the Hamming distance between the desired and actual circuit outputs.

To test this hypothesis three evolutionary runs were done using the same initial population, but with differences in the manner in which circuit function is evaluated. The first used a standard Hamming distance, while the second sums the number of consecutive matching initial signals on each output, and the third counts the maximum number of consecutive matching signals on each

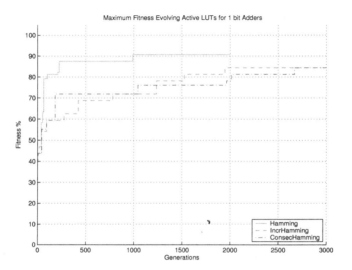

Figure 8.4: Maximum Fitness for Evolving Adders with Modified Hamming

output and sums them. All were run to stagnation, which in this case was set at 1000 generations.

The result of these runs, plotted in Figure 8.4, showed that there was no advantage, and possibly some disadvantage, provided by the non-Hamming variants. This is not entirely surprising, given that the order of signals is determined by the order of inputs applied, and doesn't contain any additional information, in regards to the functionality of the circuit's building blocks, compared to the standard Hamming distance.

The ambiguous results provided by this preliminary set of investigative experiments highlighted the need for a more comprehensive and systematic approach to locating the causes for evolution and morphogenesis failing to achieve a 100% solution.

## Preconfiguring Resources

The next attempt at isolating the causes involved in preventing evolution from finding a global optima involved starting morphogenesis from points in phenotype space that are, hopefully closer to the solution. To do this, the FPGA is preconfigured to a part of a 100% solution, based on the hand crafted solution given in Section 8.5. For each evolved individual being evaluated, morphogen-

esis is conducted from this starting point. The direct encoding GA was not
used in these experiments however, due to the fact that it performs a complete
FPGA configuration for each genotype, as compared to the morphogenetic
approach which starts with the supplied (preconfigured) FPGA layout and
incrementally applies partial reconfigurations.

Two sets of experiments were carried out for each LUT encoding. In the
first set of experiments, the LUT component of the solution needs to be found,
while the multiplexor (routing) settings have been preconfigured. In the second
set, the LUTs are preconfigured, while the multiplexor settings need to be
found. In either case, morphogenesis is free to manipulate both preconfigured
and unconfigured resources.

Two runs were done for each variant, using the same evolutionary and morphogenetic parameters as before, but with stagnation after 1000 generations.
The results of these runs are presented in Table 8.5, with experiments labeled
by LUT encoding (Bit, Incremental, Active) and the component that was not
preconfigured (LUT or Mux).

Table 8.5: Preconfigured Adder Experiment Run Results

| Experiment | Run 1 | | | | Run 2 | | | |
|---|---|---|---|---|---|---|---|---|
| | Max Fitness | Gen | Growth Steps | Genes Count | Max Fitness | Gen | Growth Steps | Gene Count |
| Bit LUT | 100.0 | 214 | 15 | 6 | 100.0 | 434 | 28 | 8 |
| Incr. LUT | 100.0 | 235 | 28 | 9 | 100.0 | 488 | 19 | 3 |
| Active LUT | 100.0 | 71 | 34 | 4 | 100.0 | 40 | 22 | 2 |
| Bit Mux | 91.25 | 1514 | 27 | 13 | 93.75 | 1417 | 52 | 5 |
| Incr. Mux | 93.75 | 2030 | 36 | 5 | 90.625 | 1972 | 47 | 10 |
| Active Mux | 94.8 | 728 | 37 | 12 | 90.625 | 993 | 16 | 10 |

Morphogenesis showed that it was able to locate a 100% solution when
provided with preconfigured routing, but not when only provided with preconfigured LUT functions. This would tend to indicate that the problem lies
with the complexity of the Virtex FPGA's routing. Further experiments that
resolve this in a more definitive manner are presented next.

Prior to this, it is worth noting that the positive results here, in which
morphogenesis was readily able to find an adder circuit when started at a
point where the LUT functions were unconfigured and routing was configured
but with both able to be modified, show that a morphogenesis process could
be used as a dynamic circuit repair mechanism. To quantify this, for Bit and
Incremental functions (which are equivalent and harder to solve than for Active
functions), morphogenesis was able to complete the circuit configuration in as
few as 15 growth steps with 6 genes (or 19 steps with 3 genes) when started at a

Hamming distance of 64 bits from the desired solution (calculated by summing the Hamming distance between the LUT settings provided in Section 8.5 and the unconfigured LUTs' setting of 1111111111111111).

### Fixing Resources

To identify where evolution fails in its task of creating a 1-bit full adder, the problem was divided into several sub-problems, each involving the generation of an adder by manipulating a subset of FPGA resources, while the rest of FPGA resources are fixed, so as to generate an adder circuit when combined with the evolving resources. The fixed resources are configured to some subset of the adder solution presented in Section 8.5.

Three sets of experiments were run, each using a different subset of resources. In the first, the LUTs' configurations are evolved while all (routing) multiplexors are fixed to the configurations given in Section 8.5. In the second set, both the LUTs' configurations and their input multiplexor settings are evolved, while all output multiplexors are fixed, as previously. In the final set of experiments, all routing (multiplexors) is evolved, while the LUT functions are fixed.

Two evolutionary runs were done each for the directly encoded GA, and for the Bit and Active LUT encodings (Incremental functions were not used, as Bit functions provided similar results in experiments to date) used with the morphogenetic approach (time constraints prevented more than this), each using the same evolutionary and morphogenetic parameters as before, with evolution being stopped at success or after 2000 generations of stagnation. The results of these are given in Table 8.6. The experiments are labeled with GA for the direct encoding, or LUT encoding (Bit, Active) for the morphogenetic approach, and by the component that was evolved (LUT, LUT In or Mux). The results from the earlier unconstrained experiments are added for comparison, and have no evolved component suffix.

This clearly shows that the problem lies in the evolution of the routing between LUTs, while LUT functions are able to be evolved quite rapidly. Further analysis of this is conducted in Chapter 9.

## 8.7    Reported Work Evolving Adders on a Virtex

At this point it should be pointed out that there has been other work evolving adders on a Virtex reported in the literature. In (Gordon and Bentley 2002a) Gordon and Bentley presented successful intrinsic gate-level evolution of 2-bit adders on a 2x2 CLB matrix using a direct encoding approach. They were, however, unsuccessful at generating 2-bit adders with a developmental

Table 8.6: Fixed Adder Experiment Run Results

| Approach | Run 1 | | Run 2 | |
|---|---|---|---|---|
| | Max Fitness | At Gen | Max Fitness | At Gen |
| GA LUT | 100.0 | 29 | 100.0 | 73 |
| Bit LUT | 100.0 | 223 | 100.0 | 33 |
| Active LUT | 100.0 | 4 | 100.0 | 0 |
| GA LUT In | 93.75 | 122 | 96.875 | 1106 |
| Bit LUT In | 93.75 | 210 | 100.0 | 2727 |
| Active LUT In | 93.75 | 98 | 100.0 | 2252 |
| GA Mux | 93.75 | 150 | 96.875 | 1616 |
| Bit Mux | 81.25 | 798 | 84.375 | 2113 |
| Active Mux | 84.375 | 757 | 90.625 | 3598 |
| GA | 96.875 | 174 | 96.875 | 536 |
| Bit | 87.5 | 1369 | 93.75 | 4128 |
| Active | 96.875 | 670 | 96.875 | 1849 |

approach (Gordon and Bentley 2002b). As with the work in this book, JBits was used to configure the FPGA.

Unlike the work presented earlier in this chapter, the CLBs on the edges of the evolvable region were constrained to use only circuit input wires and connections from other evolved cells. Each LUT on the south west edge CLBs had three of its four inputs set to A0, B0 (the low order bits to be added) and Cin, while the other input could be either be one of these or come from the CLB to the north (the other input CLB) which has two of its LUT inputs set to A1 and B1 (the high order bits to be added). The circuit's outputs were taken from the hex lines on the opposite edge (Cout and the high order sum from the north east and the low order sum from the south east CLB), which were not evolved. Again, unlike the work presented in this chapter, CLB control input muxes and internal CLB logic were also evolved. In common with the work presented in this chapter, however, the CLB output bus muxes and directly connecting single PIPs are evolved (although in their case only 40 of the 48 are evolved to avoid contention).

The chromosome encoded the 604 FPGA resources that were evolved as integers, to avoid arbitrary bias, except for the LUTs which were encoded as 16 bits (i.e. native format).

One-point crossover was used with a 100% rate of application, along with a mutation rate of 5% of all genes. Two member tournament selection was used, with a 70% probability of the winner being chosen, and the population size was set at 50.

As circuits weren't constrained to be purely combinatorial (registers were allowed), each circuit was evaluated five times with input vectors randomised for each evaluation. Circuit fitness was based on the number of correct bits in the output sequence, and the fitness assigned to the chromosome was the lowest of the five circuit evaluations.

Ten runs were conducted, with nine successfully generating a 100% solution. The successful runs required on average 2661 generations to find the solution, with a standard deviation of 2169 generations.

While Gordon and Bentley's direct encoding work was able to find a 2-bit full adder, where the direct encoding and morphogenetic approaches to unconstrained adder experiments presented in this chapter failed to find a 1-bit adder, the two problems were significantly different. In the former case, both combinatorial and sequential circuits were allowed (as compared with combinatorial only in the latter case), and possibly more importantly parts of their layout were fixed (the circuit inputs and outputs were largely fixed) and routing was reduced to ensure signal flow from inputs to outputs (with crossover between the two major signal pathways allowed, without which carry propagation would not be possible). These factors make it difficult to compare the performance between the two approaches.

## 8.8   Summary

In this chapter an investigation into the performance of both traditional and morphogenetic approaches to EHW was conducted in regards to the evolution of circuit structure *and* functionality on a more complete set of resources than was used in earlier experiments (in Chapter 6). To do this, a one bit adder circuit on a 2x2 CLB region of the FPGA, was chosen as the target for evolution.

Circuit fitness was based on progress towards connecting input and output cells and on the output of the circuit when signals are applied to the inputs. A perfect solution required the circuit to generate the correct outputs (*sum* and *carry out*) for each combination of circuit input (*x*, *y* and *carry in*) signals.

Unfortunately, neither the morphogenetic or traditional approaches to EHW were able to generate a 100% correct circuit, that being one generating the correct outputs for all combinations of inputs, however they were able to connect inputs to outputs easily, and managed to generate solutions that were close to optimal, having mostly correct output signals. Surprisingly, the traditional EHW approach outperformed its morphogenetic equivalent (with native LUT encoding), although it was surmised that this may be partly due to the higher effective mutation rate aiding exploration. Furthermore, it is also possible that within the small 2x2 CLB region there is little room for a morphogenesis process to provide much additional assistance to circuit generation.

Further experiments were undertaken to isolate and identify the causes of failure. The results of these experiments seemed to indicate that evolution (and morphogenesis) had problems searching the routing multiplexor space, while it had no difficulty in finding correct LUT configurations. These results will be further analysed in the following chapter.

Experiments in which morphogenesis was started on a partially preconfigured solution and relatively rapidly found a 100% solution, when started with 64 bits of the configuration remaining to be found, showed that the morphogenesis process may be well suited to correcting damaged circuits, in which some portion of the FPGA configuration has been corrupted, or where there is some actual hardware damage.

These experiments were also used to evaluate the utility of different LUT encodings for aiding evolutionary search, these being the native encoding (Bit), incrementally applied basic Boolean functions (Incremental) and basic Boolean functions applied only to LUT inputs carrying active signals (Active). From the experiments carried out here, it appears that Incremental functions provided no advantage over the native binary encoding, while Active functions proved to be superior in terms of speed of evaluation and evolutionary performance within the constraints of this set of experiments.

In the next chapter, the results of these experiments, and those from Chapter 6, are analysed in depth to determine when EHW problems will be solvable, by the morphogenetic approach in particular.

Chapter **9**

# Estimation of Problem Solvability

The aim of this chapter is to determine what the factors are that determine whether a given EHW problem will be solvable or not, by the morphogenetic EHW approach in particular. With these factors identified, a heuristic may be developed for identifying whether or not any given problem is likely to be solvable by the morphogenetic EHW approach.

To this end, the first section develops measures of circuit complexity, as implemented on an FPGA, and feedback from the fitness function for guiding the morphogenetic process to a solution. These measures have been applied to the experiments presented in this book (see Appendix B for details of the actual calculations), and results are presented in Section 9.2. Using these results, a heuristic, for determining whether or not the morphogenetic approach will succeed at solving a given problem, is provided in Section 9.3, along with guidelines for increasing the likelihood of success. The chapter is then summarised in Section 9.4.

## 9.1  Measuring EHW Circuit Complexity

As EHW is being undertaken within the constraints of an FPGA, the most obvious measure of circuit complexity, is the size of the evolved region. In Section 6.4.1 this approach was used for comparing the relative performance of different EHW approaches to routing signals across CLB matrices of increasing size. Another measure, which is also able to be used to compare performance when varying sets of FPGA resources are used, is the size of the configuration state space. However, neither of these measures are suitable for comparing the complexity of circuits with different structural or functional requirements, but with the same CLB matrix size and FPGA resource subset. So, this approach could not be used to compare the experiments presented in Section 6.4.2, for example.

In terms of evolutionary computation and EHW, an often used measure is that of problem difficulty. This is generally measured qualitatively in terms of the fitness landscape, and is typically performed by undertaking a series of random walks from a global optima.

Unfortunately, such a stochastic approach is unsuitable for problems that require long evaluation trials, such as the adder experiments presented in Chapter 8. Instead, what is required here, for determining why some problems were solved while others weren't, is a quantitative measurement of problem difficulty that takes into account the complexity of the evolved region of the FPGA, and a measure of the feedback supplied by the fitness function for guiding the evolutionary and morphogenetic processes to a solution. These two factors are explored in the following two subsections.

### 9.1.1   Measuring Circuit Complexity

To define a measure of complexity for EHW undertaken at the gate-level (in which the FPGA's multiplexors are individually configurable), the most obvious measure is the state space of the problem, calculated as the total number of configuration states to which the FPGA may be configured by the EHW system.

Another measure of problem complexity, is its difficulty, as defined by the probability of a random configuration of the FPGA being a solution (that is, having a fitness evaluation of 100%).

#### Cell Configuration State Space

To calculate the state space of the evolvable region of the FPGA, it is first necessary to determine the state space of a single EHW cell, which in this case is a logic element (though a CLB or slice could be used if a different cell mapping is used). The state space of this element ($M$) is defined as the product of the number of settings ($\sigma$) available for each of its $m$ constituent multiplexors and LUTs that are manipulated by the EHW system, that is

$$M = \prod_{i=1}^{m} \sigma_i. \qquad (9.1)$$

#### Native LUT Configuration State Spaces

For a LUT with a native (Bit) or incremental encoding, the LUT's configuration is defined by its Boolean truth table, and so the number of settings is defined as

$$L_b = \begin{cases} 2^{2^l} & 1 \leq l \leq 4 \\ 1 & l = 0 \end{cases} \qquad (9.2)$$

**172**

where $l$ is the number of used LUT inputs, giving $L_b$ possible values of 1, 4, 16, 256, and 65536 configurations. Note that LUTs encoded using *LUT Incremental Functions* (see Section 8.1) have the same state space as native encodings, as they can visit all possible LUT configurations.

**Active LUT Configuration State Spaces**

For LUTs encoded with *LUT Active Functions* (presented in Section 8.1), the state space is determined according to the number of basic Boolean functions of active inputs, as given by

$$L_a = \begin{cases} 1 & l = 0 \\ 8 \cdot 2^l & 1 \le l \le 4 \end{cases} \tag{9.3}$$

where 8 is the number of basic functions, $l$ represents the number of used LUT inputs, and $2^l$ factors in the number of inverted and noninverted input combinations available to the function. This gives $L_a$ possible values of 1, 16, 32, 64, and 128 configurations.

Obviously, when Active LUTs are used, the number of active inputs will vary from cell to cell, and even from one growth step to the next, meaning that the value of $L_a$ is variable, and so too is a cell's value of $M$. So, instead of relying on a constant $L_a$, an approximation will have to suffice for calculating $M$ in the general case.

Noting that the possible values of $L_a$ in base 2 logarithms is 0, 4, 5, 6, 7 for LUTs with 0-4 active inputs, and assuming a uniform distribution, then the average active LUT's function state space can be approximated as the mean of the possible values of $L_a$, giving an estimate of $2^{4.4}$ for $L_a$. However, if it is taken into account that when a LUT becomes inactive (i.e. not connected in some manner to the circuit's functionality) its function is not updated (hence, it doesn't become a function of no variables, as only active LUTs are updated, as they are the only LUTs that will have any effect anyway), then the mean of these possible values of $L_a$ (4-7), gives an estimate of $2^{5.5}$.

So, by taking a rounded average between these two estimates, a rough estimate of the Active LUT state space can be given as

$$L_a = 2^5 = 32. \tag{9.4}$$

**Evolvable Region Configuration State Space**

Once $M$ has been calculated, the total number of configuration states available to the evolvable region of the FPGA can be calculated as

$$S = M^n \tag{9.5}$$

where $n$ is the number of cells (logic elements) being evolved.

### Number of Solutions

The next measure required is the number of solutions, or answers $A$, to the problem within the constraints of this configuration space. This is problem dependent, but in all cases, elements of the configuration space that don't affect the solution (i.e. redundancies) increase the number of solutions, as

$$A = \alpha M^\rho \tag{9.6}$$

where $\alpha$ is the initially calculated number of solutions for the essential (or actively participating) circuit components (logic elements), and $\rho$ is the number of redundant logic elements. This means that the most probable solutions (or sets of related solutions), are those that require the least resources, and hence dominate the solution space. A simple example of this is in the signal routing experiments, where the shortest paths dominate.

### Problem Difficulty Definition

With the state space $(S)$ and number of solutions $(A)$ calculated, then the probability $(P)$ of a randomly generated configuration being a solution is given by

$$P = A/S \tag{9.7}$$

which can be seen to be a quantitative measure of problem difficulty.

Note, however, that this does not provide any information regarding the fitness landscape itself, which would give a more accurate measure of how hard a problem is to solve.

## 9.1.2   Measuring Fitness Feedback

The fitness function defines the search landscape, and hence the ability of evolutionary and morphogenetic approaches to solve a given problem. Thus, some measure of the feedback provided by the phenotype (the growing and evolving circuit) to the fitness function (providing a means of evaluating the phenotype, and ultimately the genotype) is essential in determining the difficulty of any problem posed to an evolutionary or morphogenetic system.

To create a quantitative measure of the fitness function, the primary measurement is the size of the fitness feedback space, this being the amount of information provided by the system to the fitness function for guiding evolution.

### Fitness Feedback Space

There are two components that contribute to the fitness feedback space, in general. The first, structure specifying, component is the feedback supplied by the circuit elements, given per cell (logic element) as $\mu$.

To clarify this, an example of this component of feedback is the connectivity evaluation provided for each cell ($\mu$) in the signal routing experiments. Each cell's connectivity is defined, by the circuit connectivity analyser, as one of five possible states, with the latter including each of the preceding (see Section 6.3 for more details). These connectivity states are: disconnected, input connected to an active signal, function passes signal, output passed to out bus, or signal passed to a connected cell. This information is supplied for each cell, and passed to the fitness function, thus giving a value of $\mu = 5$ for this problem.

The structure specifying component of feedback is calculated as

$$F_s = \mu^n \tag{9.8}$$

where $n$ is the number of cells.

The other, function specifying, component is the feedback supplied by the overall circuit function, measured (for combinatorial, stateless problems) in terms of the number of circuit inputs ($I$) and outputs ($O$) required to completely describe the circuit function. For combinatorial circuits, this would typically involve applying each possible circuit input combination, and sampling all the circuit outputs to construct a Boolean truth table.

This (combinatorial) function specifying component of feedback is calculated as

$$F_f = 2^{(2^I O)} \tag{9.9}$$

where the 2's in the equation are due to the binary nature of digital signals.

These two fitness components can be combined to give an overall fitness feedback space of

$$F = F_s F_f = \mu^n 2^{(2^I O)}. \tag{9.10}$$

**Effective Fitness Feedback Space**

While Eq. 9.10 tells us the total amount of information available to the fitness function, it doesn't take into account how much of this information is actually used to calculate the fitness. Thus, this can be further refined, to create an *effective* fitness feedback measure ($E$).

For example, in the experiments presented in this book, when connectivity is measured, not all cells' (connectivity) feedback is utilised. Specifically, in the circuit input and output layers (columns of CLBs) only the connectivity of the cells corresponding to circuit input and output logic elements is utilised, while all other cells in these layers have their connectivity measure discarded in the fitness evaluation. Hence, in the calculation of E $n$ is reduced to $n'$.

Furthermore, in some of the adder experiments where part of the circuit is fixed, such as fixing the routing muxes so as to only evolve the LUT functions for example, the amount of connectivity feedback provided by each logic element is often reduced, from $u$ to $u'$.

Combining these would reduce the earlier measure of $F$ to an *effective* fitness feedback space ($E$) of

$$E = (\mu')^{n'} F_f. \tag{9.11}$$

## 9.2   Complexity Calculation Results

Using the measures introduced in the previous section, it is possible to calculate each experiment set's characteristics, in terms of configuration state space size, fitness feedback space, and problem complexity as measured by the probability of a random FPGA configuration being a solution.

The calculation of these requires that the specifics of the FPGA resources, implementation of the experiment on the FPGA and fitness function implementation, are each taken into account. The amount of work required to determine the number of solutions (FPGA configurations) to a given problem is substantial. To present the working here would distract from the main aims of this chapter, that being to determine what the crucial features of the experiments were that prevented the majority of the adder experiments from attaining a global maximum. For this reason, this experiment specific information, and the calculations themselves, have been moved to Appendix B, while only the results of these calculations are presented here in Table 9.1. To this table has been added a few extra columns based on the results of the calculations, which will be explained in the coming sections.

Note that routing experiments are labeled as CLB cols x rows ':' circuit inputs - outputs. The fixed adder experiments (those in which some subset of resources were preconfigured to a predefined solution, and thenceforth fixed) are labeled according to what components were evolved (LUTs only, LUT and input muxes, or Muxes only) and the manner in which LUTs are encoded (Bit for native encoding, or Active for Active LUT functions).

All entries (besides the percentage solved columns) are given as base 2 logarithms, and rounded to an appropriate number of decimal places. In the last two columns, the percentage of successful runs for the morphogenetic (MG) and direct encoding (GA) approaches are presented, and as Active LUT encodings were not implemented for the direct encoding approach, these are marked with "N/A".

Due to time limitations, for each of the routing problems, only 10 runs were done, while for the adder problems, two of each variant were performed. Hence, the percentage of runs solved for the adder problems, in particular, should only be taken as indicative of trends, rather than having any statistical significance.

While it would be interesting to be able to use this data to compare and generalise the performance of the two EHW approaches, as measured by gen-

**176**

Table 9.1: Experiment Calculation Summaries in Base 2 Logs

| Experiment | S | A | P | F | E | S/F | $|P|/E$ | MG % | GA % |
|---|---|---|---|---|---|---|---|---|---|
| 5x5:1-1 | 999 | 962 | -36.9 | 232 | 144 | 4.30 | 0.26 | 100 | 100 |
| 9x9:1-1 | 3237 | 3170 | -67.2 | 752 | 590 | 4.30 | 0.11 | 100 | 100 |
| 13x13:1-1 | 6754 | 6656 | -98.5 | 1570 | 1333 | 4.30 | 0.07 | 100 | 100 |
| 17x17:1-1 | 11550 | 11420 | -129.8 | 2684 | 2373 | 4.30 | 0.05 | 100 | 100 |
| 8x8:1-1 | 2558 | 2500 | -58.2 | 594 | 450 | 4.30 | 0.13 | 100 | 100 |
| 8x8:4-4 | 2558 | 2349 | -208.6 | 594 | 464 | 4.30 | 0.45 | 100 | 100 |
| LUTBit | 256 | 231 | -25.4 | 32 | 21 | 8.00 | 1.21 | 100 | 100 |
| LUTinBit | 496 | 422 | -73.3 | 32 | 21 | 15.49 | 3.49 | 50 | 0 |
| MuxBit | 464 | 337 | -127.0 | 32 | 21 | 14.49 | 6.05 | 0 | 0 |
| AllBit | 720 | 565 | -144.5 | 32 | 21 | 22.49 | 6.88 | 0 | 0 |
| LUTAct | 38 | 18 | -20.4 | 53 | 16 | 0.72 | 1.28 | 100 | N/A |
| LUTinAct | 320 | 251 | -68.3 | 53 | 24 | 6.02 | 2.85 | 50 | N/A |
| MuxAct | 464 | 350 | -113.6 | 53 | 26 | 8.73 | 4.37 | 0 | N/A |
| AllAct | 544 | 410 | -132.8 | 53 | 28 | 10.23 | 4.81 | 0 | N/A |

erations required to solve the problem, to increasing problem difficulty, neither $P$ or $S$ is satisfactory for this, as the performance of any EHW approach is highly dependent on the fitness landscape.

For example, this data shows that for an 8x8 CLB matrix, the difference in complexity between routing a single signal and routing four signals is roughly $2^{150} \approx 10^{45}$. This required an increase of less than four times the generations for the direct encoding approach to solve (and was solved in less generations by the morphogenetic approach). On the other hand, for the single signal routing, for an increase in CLB size from 5x5 to 17x17 CLBs, this incurs considerably less increase in difficulty ($2^{93} \approx 10^{28}$), however there was a much larger increase in generations required. In this case, the better performance, relative to increase in complexity, provided by scaling IO versus scaling CLB matrix size, can be readily explained by the domain knowledge provided by the fitness function, which guides evolution towards increasingly spreading routing across the breadth of the columns as the CLB matrix is traversed.

## 9.3    Solvability Heuristic

### Determining Problem Solvability

By examining Table 9.1, it seems that dividing the magnitude of the logarithm of the probability of a random configuration of the FPGA ($|\log(P)|$) by the logarithm of the amount of feedback utilised by the fitness function ($\log(E)$), denoted as $|P|/E$ in the table and in the following paragraphs, correlates well with the solvability of the problem, by an evolutionary or morphogenetic approach to EHW. On the other hand, as expected, $\log(S)/\log(F)$, which shows the total amount of feedback available to be used to guide the morphogenetic system within the entire configuration state space, doesn't appear to correlate well.

The motivation for calculating $|P|/E$ is that $|P|$ can be seen as a way of reducing the size of the state space $(S)$, by removing the redundancies that were used to increase $A$ from the small number of possible solutions $(\alpha)$, to a size where there is only a single solution. Therefore, $|P|/E$ shows the amount of effective fitness feedback coverage of this smaller state space, which obviously maps the entire space for $|P|/E \leq 1$, while increasingly leaving gaps for $|P|/E > 1$.

This approach makes the assumption that all the solutions in the original state space are uniformly distributed, however, given that $|\log(P)|$ is derived largely through the redundancy of unused cells, this seems a reasonable assumption (i.e. the redundancy that creates the majority of the alternate solutions doesn't clump them all together in the same area of state space).

From the experiments conducted to date, it appears that for the problem to be consistently solvable, the value of $|P|/E$ needs to be less than or equal to one, with less than one providing more feedback than necessary, and as this value increases the experiment's likelihood of being solved decreases. From the limited experiments conducted, values of $|P|/E$ just below 3 had a 50% chance of being solved, those around 3.5 had around a 25% chance of being solved, and those around 4.5 or more, were unlikely to be solved (in a limited number of runs).

As can be seen, the routing experiments, while they are harder to solve, according to $P$, remain solvable due to the large amount of effective feedback to guide evolution. while, on the other hand, the unfixed adder problem and the adder problem in which multiplexor settings only need to be evolved both lack sufficient feedback to guide evolution to a full solution. It can also be seen that evolving the adder LUTs was successful due to the larger proportional coverage of feedback to the problem.

### Determining Required Effective Feedback

Given that $|P|/E$ correlates well with the solvability of the problem, and experimentally having determined (tentatively) that for a given problem to be consistently solvable requires this value to be less than or equal to $k$, where $1.28 \leq k < 2.85$, then $k$ can be used to determine how much effective feedback is required to solve a given problem. For simplicity, this range is replaced with a single value of $k = 2$, for determining how much effective feedback would be required for the problem to solvable most of the time (remembering that at

Table 9.2: Adder Experiment calculated E and required E* ($\log_2$), for k=2

| Experiment | P | $|P|/E$ | E | $E^*$ | Bits Required |
|---|---|---|---|---|---|
| LUTBit | -25.415 | 1.210 | 21.000 | 12.708 | 0 |
| LUTinBit | -73.292 | 3.490 | 21.000 | 36.646 | 15.646 |
| MuxBit | -127.044 | 6.050 | 21.000 | 63.522 | 42.522 |
| Bit | -144.459 | 6.879 | 21.000 | 72.230 | 51.230 |
| LUTAct | -20.415 | 1.276 | 16.000 | 10.208 | 0 |
| LUTinAct | -68.292 | 2.854 | 23.925 | 34.146 | 10.221 |
| MuxAct | -113.569 | 4.368 | 26.000 | 56.785 | 30.785 |
| Act | -132.814 | 4.810 | 27.610 | 66.407 | 38.797 |

$k = 2.85$ 50% of the time a 100% solution was found).

Thus, remembering that $P$ is always negative (by definition of probability), it is required that

$$- \log_2(P)/\log_2(E) \leq k \qquad (9.12)$$

rearranging this gives

$$\log_2(E) \geq - \log_2(P)/k \qquad (9.13)$$

then moving the terms on the right hand side within the log operator

$$\log_2(E) \geq \log_2((1/P)^{1/k}) \qquad (9.14)$$

and as $\log_2$ is a monotonic function, it can be removed from both sides while preserving the relation, thus

$$E \geq (1/P)^{1/k}. \qquad (9.15)$$

Denoting $(1/P)^{1/k}$ as $E^*$, gives the minimum amount of effective feedback (in $\log_2$, i.e. bits) required to guide the morphogenetic system to a solution. Hence, for a given problem to be solvable, it is required that $E \geq E^*$.

The results of the calculations of $E^*$ for the adder experiments, with $k = 2$, are presented in Table 9.2.

### Increasing Likelyhood of Success

From Table B.5 in Appendix B.2.2, it is known that of the 8 utilised logic elements in the partially fixed adder experiments, only 5 (corresponding to the assigned IO cells) have their connectivity measure taken into account, that is $n' = 5$. If $n'$ is increased by 3 to utilise the 8 used logic elements, then given that $\mu' = 2$ for Bit encoded LUTs, while $\mu' = 3$ with Active LUT encodings for evolving LUTs and their inputs, and $\mu' = 4$ for evolving muxes only, it is possible to determine if these experiments could have been successful (using

Eq. 9.11) if more connectivity feedback was utilised by the fitness function.

For the Bit encoded LUTs, this would provide an extra 3 bits of effective feedback, while for Active encoded LUT approaches, this would give 4.755 extra bits for evolving LUTs with their inputs, and 6 bits for evolving muxes. In all cases, too few to ensure success.

For evolving the unconstrained adder, all 16 logic elements can be utilised, and so $n'$ can be increased from 5 to 16. Given that $\mu' = 2$ for Bit encoded LUTs, and $\mu' = 5$ with Active LUT encodings for the unconstrained experiments (again from Table B.5), then it can be seen that the Bit encoded problem gains an extra 11 bits of feedback, while the Active function encoding gains 25.541 bits. In both cases, this falls short of the required amount of feedback (by 40 and 13.256 bits respectively) to ensure success.

From this it can be seen that these problems would still have lacked sufficient effective feedback to ensure that evolution would be guided to finding a 100% solution. So, with $n'$ increased as far as possible, and with the functional component of feedback already utilising all its available information (i.e. $\mu'$ can't be increased), the only way to increase the likelihood of success, without changing the general approach to fitness evaluation, would be to increase the amount of structural feedback. That is $\mu$ needs to be increased somehow, which would hopefully also result in $\mu'$ being increased.

If such an approach is still unable to solve the adder problem, then more functional feedback would need to be generated. This would require evaluating the functional aspects of the logic elements within the circuit path, to ascertain how they contribute to circuit function, and hence to determine the utility of the function they perform in the context of the entire circuit.

## 9.4   Summary

In this chapter, quantitative measures of problem difficulty, in terms of the probability of a random FPGA configuration being a solution, and the amount of fitness feedback information available for guiding evolution and morphogenetic processes, have been introduced.

Based on these, it is apparent that the determining factor in whether a given problem is solvable or not is the amount of feedback available relative to the problem difficulty. That is, by assuming a uniform distribution of solutions (which is not unreasonable due to the manner in which circuit component configuration redundancies dominate the solution spaces) and reducing the size of the state space to only include a single solution, it is necessary that sufficient coverage of this space is provided by the fitness function for it to be able to search and locate the solution. Empirically, it has been found that complete coverage appears to guarantee success, while with decreasing coverage, the probability of failure increases.

By using this heuristic, a problem's tractability can be ascertained, and if, according to this measure, a given problem appears unsolvable, then either more feedback is required, or the problem difficulty needs to be reduced.

In conclusion, while problem difficulty and state space size may influence the number of generations required to solve a given problem, it is the amount of useful fitness feedback information supplied to the EHW system that determines whether or not a given problem can be solved by evolutionary and morphogenetic methods.

Chapter **10**

# Conclusion

This chapter ties together the work presented in previous chapters, to weigh up the results of this work in light of the original aims, and to provide directions for future work.

First, Section 10.1 presents a summary of the outcomes of each the previous chapters, then in Section 10.2 the results of this work are evaluated to identify contributions to the field of evolvable hardware. The chapter is then concluded in Section 10.3 with a discussion of some of the issues that have arisen during this work, and outlines avenues of future work based on these.

## 10.1 Summary

### Chapter 1: Introduction

Gate-level evolvable hardware (EHW) has been introduced, in Chapter 1, as an alternate design method that is not constrained by human design preconceptions, and so better able to utilise the available hardware resources. EHW also has properties of adaptation and fault tolerance that make it an attractive technology for future space exploration applications. These properties also make EHW well suited to the autonomous robotics domain, which was the original inspiration for the work undertaken in this book.

One of the primary factors preventing EHW from achieving more widespread use is the issue of scalability — traditional approaches to EHW don't scale well to increases in circuit complexity. This problem has been exacerbated through the increased complexity of modern commercially available field programmable gate arrays (FPGAs).

Abstracting away the architectural details of the target device has become a common approach to dealing with this, but results in the loss of many of the properties of EHW that made it an attractive research field. Alternatively, encoding a growth process (known as morphogenesis) on the chromosome has been proposed as a means of scaling EHW without requiring architectural details to be abstracted away.

With this in mind, the central aim of this work was defined to be scaling EHW to increases in circuit size and complexity on commercially available reconfigurable hardware without requiring architectural details to be abstracted away (i.e. gate-level EHW on a commercial FPGA) by developing a morphogenetic approach to EHW that utilises the hardware architecture. Solving this problem could allow gate-level EHW to once again be successfully applied by researchers to a wide variety of difficult engineering problems using commercial-off-the-shelf hardware.

### Chapter 2: Evolvable Hardware Background

In Chapter 2 EHW was defined, and EHW platforms, applications and approaches were surveyed. From this survey, it is apparent that gate-level EHW offers the most value, being able to fully utilise its underlying medium, while the scalability problem is identified as a major obstacle to progress in EHW.

The Xilinx Virtex FPGA was identified as the only suitable commercially available platform for gate-level EHW. The Virtex allows gate-level access to its resources through the JBits Java API. However, this introduces another problem, not seen in earlier gate-level EHW work on the EHW-friendly Xilinx 6200 FPGA, in that it is possible to configure the FPGA to potentially damaging configurations.

Some researchers have suggested that the solution to the scalability problem is to encode a growth process, known as morphogenesis, on the evolved chromosomes, rather than directly encoding the FPGA configuration.

### Chapter 3: Morphogenesis Background

Chapter 3 covered the properties of morphogenesis that make it attractive for scaling gate-level evolvable hardware. While the morphogenesis process itself offers a great deal of scalability, this is further enhanced through a high level of redundancy in the genetic encoding which provides neutral pathways through evolutionary space, and robustness to genetic operations.

A review of morphogenesis in EHW and evolutionary computation was undertaken, from which it was apparent that cell-based models driven by gene expression offer the most promise for evolving circuits within the constraints of an FPGA.

### Chapter 4: Designing a Morphogenetic EHW System

Chapter 4 developed a morphogenetic model for EHW based on the processes and mechanisms from Chapter 3 that can usefully be applied to gate-level

EHW on an FPGA. Morphogenesis driven by gene expression was chosen as the primary biological abstractions applicable to this EHW model.

A close hardware-biological model mapping was decided on, which avoids arbitrary design decisions. EHW cells are mapped to logic elements, due to their functional independence, while the configuration of the multiplexors and LUTs within the logic element are represented as proteins within the cells, and inter-cellular signaling pathways are provided by the routing wires.

A transcription-level gene regulation model was chosen, loosely based on gene regulation in prokaryotes, in which genes are encoded on a variable length base-4 chromosome with their components identified by signatures, providing both flexibility and redundancy.

Genes encode proteins, and when these are transcribed by polymerase, they are able to change the cell's gate-level FPGA settings. These genes are in turn regulated by proteins, representing the cell's current gate-level settings. Regulation is achieved through the binding of proteins to regulatory regions of the gene. Binding to regulatory regions upstream of the promoter site enhance the binding of polymerase to the promoter, and hence the likelihood of transcription occurring, while binding to regulatory regions between the promoter and the coding region repress the binding of polymerase to the promoter, and hence lower the likelihood of gene transcription.

A redundant genetic code is used for mapping from codons (triplets of bases in the gene's coding region) to intermediate gene products, which are in turn decoded to the JBits class and values required to configure the FPGA resources. A related (codon-based) bind code is used to determine the signature sequences for FPGA resource settings. A template-matching approach is used for determining binding of proteins to gene bind sites.

Biologically inspired genetic operators were designed to support the chromosome encoding. These include homologous crossover, which performs crossover at regions of high similarity, thus reducing the likelihood of genetic damage, and mutation operators that were designed to take advantage of redundancy in the genetic and bind codes to provide neutral mutations.

### Chapter 5: Morphogenetic EHW Implementation

Chapter 5 provides details of the implementation of the gate-level morphogenetic EHW (MGEHW) system for the Xilinx Virtex FPGA, based on the design given in Chapter 4. This system is made up of a collection of processes that perform layout of the FPGA, generate a population of chromosomes, and evolve this population to completion. Chromosomes are parsed and decoded (with the genetic code presented in Appendix C) and then the extracted and decoded (to JBits specification) genes are passed to the circuit evaluator, which instantiates and iterates the morphogenesis process. Interactions between genes and the FPGA result in the generation of a circuit, which is assigned a fitness metric based on its performance at a given task. The phenotype's fitness is returned to the evolutionary process, which uses this to determine which chromosomes should be selected for breeding or removal.

### Chapter 6: Experiments with Evolving Circuit Structure

With a working morphogenetic EHW system in place, Chapter 6 proceeded to evaluate and compare the performance of the morphogenetic and traditional direct encoding EHW approaches to increases in circuit size and structural complexity.

The experiments involved routing signals from designated input points on one side of a matrix of CLBs to designated output locations on the opposite side, with severely constrained routing to prevent direct horizontal paths across the CLB matrix.

Four sets of experiments were conducted in which a single signal is routed across CLB matrices of increasing size. The number of logic elements evolved ranged from 100 on the 5x5 matrix, through to 1156 on the 17x17 CLB matrix. From these experiments it was evident that with increasing size, the direct encoding approach required increasingly longer to find a solution, while the morphogenetic approach was relatively constant in the number of generations required to solve the problem, and in all cases required less generations on average to find a solution than the direct encoding approach.

An additional two sets of experiments were conducted on a mid-sized CLB matrix containing 256 cells. The first involved routing a single signal across the matrix, while in the second four signals were routed from the center rows across to points spread across the entire edge of the opposite side of the matrix. In this case, the direct encoding approach required roughly four times as many generations to solve the multi-signal problem as the single-signal problem (possibly corresponding to the fourfold increase in IO), while the morphogenetic approach required no extra generations for the more complex problem. Additionally the morphogenetic approach solved both problems (on average) in less generations than was required by the direct encoding approach for the single signal problem.

In all experiments, the length of chromosome for the direct encoding approach was proportional to the size of the evolved CLB region, while the morphogenetic approach had no direct correspondence between CLB size or circuit complexity and chromosome length or gene count.

These experiments clearly showed the superior performance of the morphogenetic approach to EHW, and its ability to scale EHW to increases in circuit size and structural complexity.

The results of initial exploratory experiments, speed tests and sensitivity analysis are documented in Appendix A. The exploratory experiments demonstrated similar performance, further backing the experiment results documented here.

## Chapter 7: Simulated Secondary Developmental Mechanisms

In Chapter 7 the morphogenetic model was enhanced through the addition of simulated molecules which have no correspondence with the underlying hardware, to support cell-specific differentiation, axis specification and chemical gradients, all of which provide location specific information to the developmental process. Another addition was non-coding mRNA, which has been found to add an extra layer of gene regulation in higher-level organisms.

The performance of the morphogenetic EHW system with and without simulated molecules was compared on evolving circuit structures. Performance was measured on several of the signal routing experiments that were used in the previous chapter (5x5 CLBs with single input and output, 8x8 CLBs with four inputs and four outputs, and 13x13 CLBs with a single input and output). Results showed a slight deterioration in performance in all cases, and also incurred extra computational overhead.

Several factors were identified that may have contributed to this lack of performance: the arbitrariness of bind sequences and molecule placement, the choice of parameters for simulated molecules, the implementation of morphogen gradients, and the evolutionary cost of an additional layer of genetic regulation. Each of these require a great deal of testing, in isolation, to determine what their actual effects are.

The choice of parameters used here was made based on the results of the exploratory experiments presented in Appendix A, which provided similar results.

## Chapter 8: Experiments with Evolving Circuit Functionality

Chapter 8 investigated the evolution of circuit structure and functionality on a more complete set of FPGA resources by both the morphogenetic (without the addition of simulated molecules) and direct encoding EHW approaches.

The target circuit was a one bit adder on a 2x2 CLB region of the FPGA. Circuit fitness was based equally on progress towards connecting input and output cells and on the output of the circuit when signals are applied to the inputs. A perfect solution required the circuit to generate the correct outputs (*sum* and *carry out*) for each combination of circuit input (*x*, *y* and *carry in*) signals.

Unfortunately, neither the morphogenetic or traditional approaches to EHW were able to generate a 100% correct circuit, that being one generating the correct outputs for all combinations of inputs, however they were able to connect inputs to outputs easily, and managed to generate solutions that had mostly correct output signals. Unlike the previous experiments, the traditional EHW approach outperformed its morphogenetic equivalent (with native LUT encoding), although it was surmised that this was due to the higher effective mutation rate aiding exploration. It is also possible that in a small region (4 CLBs in this case) there is little room for a morphogenesis process to provide much additional assistance to circuit generation.

Further experiments were undertaken to isolate and identify the causes of failure. The results of these experiments seemed to indicate that evolution had problems searching the routing multiplexor space, while it had no difficulty in finding correct LUT configurations.

Experiments in which morphogenesis was started on a partially preconfigured solution and relatively rapidly found a 100% solution, when started with 64 bits of the configuration remaining to be found, showed that the morphogenesis process may be well suited to correcting damaged circuits, in which some portion of the FPGA configuration has been corrupted, or where there is some actual hardware damage.

These experiments were also used to evaluate the utility of different LUT encodings for aiding evolutionary search. The LUT encodings investigated were the native encoding (Bit), incrementally applied basic Boolean functions (Incremental) and basic Boolean functions applied only to LUT inputs carrying active signals (Active). From the experiments carried out here, it appears that Incremental functions provided no advantage over the native binary encoding, while Active functions proved to be superior in terms of speed of evaluation and evolutionary performance within the constraints of this set of experiments.

### Chapter 9: Estimation of Problem Solvability

Chapter 9 provides quantitative measures of the difficulty of the experiments presented in Chapters 6 and 8. This is measured in terms of the probability of a random FPGA configuration being a solution, and the amount of fitness feedback information available for guiding evolution and morphogenetic processes.

This was used to determine what factors determine whether a given EHW problem will be solvable (by the morphogenetic EHW approach in particular). From this it appears that while problem complexity and state space size may influence the number of generations required to solve a given problem, it is the amount of useful fitness feedback information provided to the EHW system that determines whether or not a given problem can be solved by evolutionary and morphogenetic methods.

A heuristic, based on problem difficulty and the amount of useful feedback supplied to the evolutionary (and morphogenetic) system, was developed for identifying whether or not any given problem is likely to be solvable by the morphogenetic EHW approach.

The calculations on which this work is based are found in Appendix B.

## 10.2    Project Outcomes

### Original Aims

The original aims stated in the introduction were to scale EHW to increases in circuit size and complexity on *commercially available* reconfigurable hardware without requiring architectural details to be abstracted away.

### Proposed Method

The method that was proposed to achieve this was to use a morphogenesis process that is closely tied to the reconfigurable hardware's architecture, and is supported by a genetic representation and operators that allow this process to be encoded and manipulated by evolution in a manner that assists the search for circuits matching the desired criteria.

### Working Morphogenetic EHW Model

At the commencement of this work, it was unknown whether it would be possible to generate circuits on a modern commercially available FPGA with a morphogenetic approach to EHW.

The work presented in the first half of this book resulted in the development of a working biologically inspired morphogenetic system for evolvable hardware on the popular, commercially available Xilinx Virtex series of FPGAs, using the JBits API to access the FPGA's gate-level configuration.

This system is based on a model of multi-cellular morphogenesis closely tied to the gate-level architecture of the FPGA: cells correspond to functionally independent FPGA cells, proteins to the configuration state of multiplexors and LUTs (or other gate-level resources if available), and inter-cellular coordination is achieved through routing lines.

The morphogenesis process is driven by a transcription-level model of gene expression on a variable-length chromosome, supported by bio-inspired genetic operators and a redundant genetic code, to provide a high level of flexibility and encourage evolutionary exploration. Genes, their coding (post condition) and regulatory (precondition) regions, are identified by specialised sequences, while interactions between the FPGA gate-level configuration (proteins) and genes are determined by protein signatures (a bind code related to the genetic code). Together, this gives evolution the freedom to determine the number of genes, along with the number of pre- and post-conditions for each gene, rather than requiring arbitrary decisions to be made by the designer.

This system provides an extensible and flexible system that allows experimental research to be undertaken in EHW and morphogenesis, both for evolvable hardware on a Virtex FPGA, and for examining the behaviours of an artificial genetic regulatory network embedded in an artificial organism.

**Evaluation of Morphogenetic Model for Scaling Circuits**

To evaluate the performance of the morphogenetic model compared with a traditional direct encoding EHW approach to increases in circuit size and structural complexity, both approaches were applied to generating primarily structure-based circuits on a Xilinx Virtex.

From these experiments it appeared that increasing the circuit size (measured by the number of logic elements evolved) had minimal effect on the number of generations required by the morphogenetic approach. With a roughly 12-fold increase in cells, only a doubling in the average number of generations required was seen (and this was almost within a single standard deviation from the mean of the smallest circuit) by the morphogenetic approach. On the other hand, the direct encoding approach required roughly a 38-fold increase in the number of generations required to generate these circuits.

The results of two sets of experiments on the same circuit size but with a four-fold increase in the structural complexity (measured by the number of IO points to be connected) showed little difference in performance between the two circuits for the morphogenetic approach, while the direct encoding approach required on average a four-fold increase in the number of generations to find a solution.

These results support the hypothesis that a morphogenetic approach is able to scale gate-level EHW on a mainstream FPGA to increases in circuit size and structural complexity without requiring architectural details to be abstracted away. Furthermore, in all cases, the morphogenetic approach outperformed (on average) the traditional approach.

**Managing Complexity of Modern FPGAs**

The other aspect of scaling complexity in EHW, is managing the increasingly complex architectures of modern FPGAs. Not only are there more resources available on FPGAs, the actual FPGA architectures are more complex. A comparison of the simplicity of the EHW-friendly Xilinx 6200 series FPGAs (illustrated in Fig. 2.2 on page 29) and the Xilinx Virtex series FPGAs (see Fig. 2.5 on page 38 and Fig. 2.6 on page 39 for example) shows this clearly.

Both traditional and morphogenetic EHW approaches are able to manage the gate-level complexity of the Virtex to some degree by limiting the subset of resources utilised by evolution (and morphogenesis). However, the real test is the evolution of circuits that utilise a larger subset of FPGA resources than was done in the primarily structure-based experiments. Unfortunately, the experiments attempted at evolving a one-bit full adder circuit on a more complete set of resources were inconclusive on this matter: both approaches were equally unsuccessful within the limited experimental runs conducted. When the resources were further limited, it showed that LUT functions were trivial to evolve, while the complex routing was difficult, *within the constraints of this particular problem*.

### Identifying and Quantifying Feedback Requirement

The most important result coming from the adder experiments, in combination with the analysis presented in Chapter 9, was identifying the importance of feedback for guiding the EHW system. Essentially, as the magnitude of the logarithm of the problem difficulty increases, the logarithm of the amount of effective feedback required to consistently guide the EHW system to a solution increases proportionally. The results of this work provides insights that are invaluable for guiding further work in gate-level EHW on modern FPGAs.

### Resolution of Circuit Validity Problem

These experiments demonstrated that the simple contention avoidance scheme outlined in Section 5.5 was effective in solving the circuit validity problems presented in 2.5.1, and meant that all routing lines could be used without threat of damage to the FPGA by evolution.

## 10.3    Discussion and Further Work

### 10.3.1    Platform Issues in Adder Experiments

For the experiments presented in Chapter 8 the *VirtexDS* simulator that is provided with JBits 2.8 was used, with a simulated Virtex XCV50 and a speed grade of 6. A simulator was used due to the unavailability of a dedicated FPGA development platform during the development of the morphogenetic EHW system, and the need to run several experiments concurrently (at least 3 dedicated PCs were used for the adder experiments) due to time constraints.

### Speed Issues

Unfortunately the VirtexDS suffers from some problems which adversely effect the ability to perform the adder experiments. The most obvious of these is that, the simulator is slower than real hardware. It takes approximately 1.6 seconds, on a 2.8 GHz Pentium 4 PC with 496Mb RAM, to test each generated circuit (which is done every growth step). This involves a partial reconfiguration of the FPGA, applying circuit stimuli and sampling the circuit's response. Coupled with the full reconfiguration of the FPGA for each phenotype, and other morphogenetic and evolutionary overheads, this results in a throughput of around 10 generations per hour for the morphogenetic approach using Bit or Incremental function LUT encodings, up to 15 generations per hour for Active function LUT encodings used with the morphogenetic approach (circuit connectivity is evaluated without requiring circuit simulation and no signals are applied to the circuit when there are no connected outputs), and 20 generations per hour for the direct encoding approach (in which no morphogenesis is performed).

This had the effect of limiting the number of runs that could be performed for each experiment variation. Some speedup would be possible by using an FPGA development platform rather than a simulator (roughly a halving of the time required to evaluate each generated circuit), however one of the main reasons for the slowness of the morphogenetic EHW system (even if circuit evaluation is done on real hardware) is the implementation of direct interaction between the gate-level FPGA configuration (using JBits) and the decoded chromosome (the genes). This approach has a drawback in that the updated partial bitstream needs to be downloaded to the FPGA board at each growth step prior to testing the circuit's behaviour. On top of this, a full FPGA reconfiguration is done for each genotype (prior to commencing morphogenesis in the morphogenetic approach).

Another issue is the inefficient implementation of the evolutionary algorithm and chromosome decoding, which are disk-bound due to unfortunate early implementation decisions. While the morphogenesis and circuit evalution processes are implemented in Java (essentially meaning that they fully reside in memory during execution), the evolutionary process and chromosome decoding involve a large number of Perl scripts that interact through pipelines and temporary files (which results in slow execution times). The positive side of this manner of implementation is that it is robust to power failures and other computer problems that often occurred during the long experiment running times.

Future work would require that the system is re-implemented in a more efficient manner.

**Simulator Faults**

Another problem is that some circuits crash the simulator when signals are applied, or worse yet, cause it to hang indefinitely. A work around was used in the experiments presented in this book to allow evolution to continue. In the case of the simulator hanging, the circuit evaluation process spawns a thread to perform the tasks that are able hang the simulator (downloading the bitstream, applying signals to the circuit input pins, and incrementing the simulator board clock), which is killed if it doesn't complete within a given time. The circuit evaluation process then exits with an error. When the circuit evaluation fails, the error recovery code assigns any chromosome that crashed or hung a fitness of zero. This obviously can have adverse effects on the ability of evolution to discover a 100% solution.

Some circuit configurations which cause the VirtexDS simulator to hang are easily recognised, one of these being recurrent unregistered connections. When the simulator is run without timing mode (timing mode assigns a speed grade to the simulated FPGA) then even recurrent connections that occur with intervening CLBs, cause the simulator to hang when it tries to resolve the signal value. Alternatively, when timing mode is used, as was done in the experiments here (a speed grade of 6 was used for the simulated Virtex XCV50), only recurrent connections to a CLB input from its unregistered outputs are unresolvable. This problem was solved by disallowing recurrent slice inputs. However other configurations with no obvious issues may also cause the simulator to hang or crash.

This means that future work should utilise a proper FPGA development board, rather than relying on the slow and buggy VirtexDS simulator included with JBits.

## 10.3.2    Issues in Morphogenetic Scalability

### Morphogenesis vs Chromosome Encoding Contribution

While the experiments in Chapter 6 demonstrated the ability of the morphogenetic EHW to scale to increases in circuit size and structural complexity, there is currently no way of determining how much of this performance is attributable to the morphogenesis process itself and how much to the manner in which this process is encoded on the chromosome and manipulated by the genetic operators during evolution.

The variable-length chromosome in base-4 provides a high degree of redundancy, flexibility with the support of bio-inspired genetic operators. It allows for junk DNA, neutral mutations, variable numbers of genes, gene products and regulators and many other features which may aid evolutionary search.

It would be informative to isolate the performance of the morphogenesis process itself from its encoding. Experiments to facilitate this would require starting with standard genetic operators and fixed length binary chromosomes, using a set number genes (probably the experiment's existing gene ceiling) and gene pre- and post-conditions encoded at specific locations. The performance of this encoding could then be compared with the full morphogenetic EHW system performance, and incrementally aspects of the morphogenetic encoding (and supporting genetic operators) reintroduced to determine the relative contribution of these to the system's performance. This would provide valuable information on the utility of many of the individual components of the morphogenetic EHW approach.

### Scaling Size with Non-neighbour Inter-cellular Communication

Experiments with the existing model of morphogenesis showed that a morphogenesis process, based solely on gene expression and inter-cellular communication between neighbouring cells with connecting lines, was sufficient for generating circuit structures with a relatively high number of cells (up to 1156 cells on a 17x17 CLB matrix were tested). As the number of cells increases further, it is likely that at some stage longer-distance coordination will be required. This could still be achieved with only morphogenesis based on FPGA configuration, through the use of hex lines and FPGA row and column spanning long lines.

### Scaling Structural Complexity with Simulated TFs

On the other hand, as the size and complexity of circuit structure is increased, it is likely that at some stage a morphogenesis process based on hardware configuration alone will not be sufficient. At some point pattern formation and cell-specific differentiation (see Section 7.1 for details) would become valuable additions. These would require the use of the simulated transcription factors (TFs) introduced in Chapter 7. However, currently these mechanisms are ineffectual.

The first step towards resolving this would be a more thorough sensitivity analysis of the simulated TF parameters. Further investigation into isolating the effect that each of the simulated TF types has on development would also be necessary.

### Choice of Cytoplasmic Determinant Bind Sequences

For cytoplasmic determinants, which currently have arbitrarily chosen bind sequences, further investigation is required to determine how they should be chosen and placed, so as to aid rather than hinder development.

### Effects of Additional Levels of Gene Regulation

Furthermore, the addition of simulated TFs also adds another level of gene regulation, and this in turn adds an overhead in terms of the amount of evolutionary learning required to utilise these. However, with increased circuit size and complexity, and longer evolution, these additions may be able to outperform the simpler hardware-state only model. To ascertain this one way or the other would require considerable further experimentation, but would be an avenue of research that could return valuable results.

### Morphogen Concentration Models and TF Binding

Morphogens provide gene regulation based on their concentration, which is a function of the morphogen's distance from the originating cell. In the current morphogenetic EHW model, morphogen strength is implemented as a probabilistically decreasing likelihood of binding, through the concatenation of a random string of length proportional to the distance from the source cell. This approach was taken due to the exact template-matching model used for binding to gene regulatory sites. While the exact template-matching approach appears to work well for gene regulation by FPGA-based resources, which are limited to a single instance per cell, it may be not be the most optimal manner of implementing morphogens and other simulated TFs, which have multiple instances within a cell. It is possible that these would be better implemented as concentrations. Further experimentation would be required to determine the best models for simulated TFs and what approach should be used to determining the binding of simulated TFs to the chromosome.

### Enhancing the Developmental Model with Cell Growth and Death

In the current model of morphogenesis, a fixed number of cells with a fixed mapping from cells to logic elements has been chosen, along with a fixed allocation of circuit inputs and outputs. However, implementing cell growth, through cell division, and cell death may possibly give better results.

In this model, development starts with a single cell (mapped to a logic element) with its inputs being the circuit inputs, and its outputs as the circuit outputs. In the case of there being more inputs or outputs to the cell than provided by the underlying hardware, then development could start with as many cells as required to provide the necessary resources. As development progresses, cells would divide to fill the adjacent logic element, or die to leave an unused logic element (that can later be re-used). Cell division and death may be the result of genetically encoded instructions, environmental interactions (i.e. a fitness or connectivity metric), or both.

One example of where cell division and death could be useful is in conjunction with incremental evolution, whereby simple functional components are evolved first, and then evolution is allowed to continue to combine and modify these to produce a circuit with the desired behaviour.

With cell division implemented, axis specifying cytoplasmic determinants and morphogen gradients may increase the performance of the morphogenetic system, in contrast with the experiments to date where they added no advantage. This would remove the need to arbitrarily place the cytoplasmic determinants, as developmental (and evolutionary) mechanisms could be used to determine their manner of distribution, as is the case in nature.

### Incremental Evolution

Although it was posited as an advantage of a morphogenetic approach to EHW, no tests have yet been done to verify if modular development can be utilised in simulated evolution to support incremental evolution, although recent work by Gordon and Bentley (Gordon and Bentley 2005) appears to support this notion. Furthermore, determining which developmental processes are required to support modular development would be valuable future work.

### 10.3.3    Future Work on the Solvability Heuristic

In Chapter 9 a solvability heuristic was developed based on the amount of feedback utilised by the fitness function and the difficulty of the problem. This heuristic was based on the results of experiments presented in this book. However, these experiments all used the same morphogenetic and evolutionary parameters. Future work, requiring further experimentation, involving the inclusion of these parameters in the determination of a problem's solvability would be a valuable addition. Another area where this work could be expanded is in analysing previous work, presented in the literature, that evolved circuits on the Xilinx 6200 series of FPGAs. This would provide useful insights into how FPGA architecture affects the amount of feedback information required to guide evolution to solutions, and could also be used to aid in the development of more EHW (and morphogenesis) friendly reconfigurable hardware.

### 10.3.4    Hardware Implementation

An obvious improvement that could be made to the existing morphogenetic process is a speed increase through hardware implementation. Morphogenesis is implicitly parallel, and could gain huge speed increases by implementing it in hardware. The evolutionary process is considerably less parallel, and so would gain less by implementing it in hardware.

To implement morphogenesis on chip would require access to the FPGA bitstream and knowledge of the bitstream format, which is not possible for the Virtex family. Alternatively, a configurable virtual architecture would need to be built on top of the existing architecture, however this would mean that real gate-level evolution is not possible. Another alternative for speeding up the morphogenesis process would be through the use of a PicoJava processor core. This could use the existing morphogenetic system, which accesses the bitstream via the JBits API.

As has been mentioned elsewhere, current FPGA designs are not EHW friendly, neither are they morphogenesis friendly. The most optimal solution would be to design or utilise an existing morphogenetic chip with a simple regular architecture that allows on chip self-reconfiguration. Some work has already been done in this area ((Macias 1999a; Moreno, Madrenas, Cabestany, Canto, Kielbik, Faura, and Insenser 1999; Tyrell, Sanchez, Floreano, Tempesti, Mange, Moreno, Rosenberg, and Villa 2003).

With a suitable architecture, morphogenesis offers real potential for providing online autonomous and robust circuit adaptation and self-repair. By evolving genotypes for both performance and adaptivity, when a circuit ceases to fulfil its task, morphogenesis can be continued until a new circuit forms that performs as required. This could be much faster than having to restart evolution to find a new circuit, as is the current approach taken (see (Stoica, Keymeulen, Arslan, Duong, Zebulum, Ferguson, and Guo 2004) for an example of this).

In a morphogenetic approach, evolution only needs to be restarted when environmental changes are beyond the capabilities of the individual. This could provide both rapid adaptation through morphogenesis and long term, and extreme adaptation through evolution.

Currently there are only two viable approaches to autonomous learning, reinforcement learning and evolutionary computation. The first of these is limited to learning problems with small, generally discrete, state spaces, while the latter is the approach of traditional evolvable hardware. The morphogenetic approach presented in this book may provide a third approach, combining elements of both approaches to provide a new approach to adaptation and learning for electronic hardware and robotics.

Appendix **A**

# Exploratory Experiments

This chapter presents the results of exploratory experiments with the morphogenetic EHW system that were used to determine appropriate system parameters, especially for the simulated molecules, and gain insight into the working of the morphogenesis system, and the evolution of chromosome structure during experimental runs.

The rest of this chapter presents the results of the exploratory signal routing experiments, noting that these use the same fitness function and parameters as the routing experiments in Chapters 6 and 7, except where indicated.

The results of experiments comparing a traditional direct encoding approach to the morphogenetic EHW approach, based solely on hardware state without the use of simulated molecules, are presented first in Section A.1. An in depth comparison, including chromosome analysis, is then done between the two sets of morphogenetic runs in Section A.2. Then in Section A.3 simulated molecules are added to the morphogenetic system, and appropriate parameters for these are ascertained, before a comparison is made between running the morphogenetic system with and without simulated molecules.

In Section A.4 speed tests are conducted, followed in Section A.5 by a sensitivity analysis of the parameters used by evolution and morphogenesis, and lastly this section is concluded in Section A.6 with a discussion of the results of these experiments.

Note that the following abbreviations are used throughout the following sections: *5x5* and *8x8* refer to the two sets of experiments, referring to the size of the CLB matrix that is used in that particular experiment; *MG*, *GA*, and *TF* in the context of an experiment set (e.g. GA8x8, TF5x5, ..) refer to the approach used - GA is used for the direct encoding approach, MG for morphogenesis and TF for morphogenesis with simulated transcription factors (TFs).

### Differences from Experiments in Body of Book

Since these experiments were completed, several program errors in the system were found that affected its performance. The most obvious of these, in comparing the experiments presented here and those in Chapters 6 and 7, is that there were flaws in the connectivity evaluation of a few routing lines. This meant that routing problems were considerably harder to solve, as some correct solutions would not be recognised as such. This affected both the morphogenetic and direct encoding approaches equally, and so the results of these experiments is still valid as a comparison between the two approaches, highlighting the superior performance of the morphogenetic approach.

Another program error that was found only affected the morphogenesis approach. In this case, for the cells in slice one, signal pathways corresponding to the LUT inputs weren't working, hence these cells relied solely on their output single muxes for inter-cellular communication.

These program errors have since been rectified, and updated experiments are presented in the body of the book.

## A.1    Morphogenesis vs Direct Encoding

### MG vs GA Single Signal Routing on a 5x5 CLB Matrix

The first set of experiments involved routing on a 5x5 CLB matrix containing 100 cells, from an input LE cell (7,9,S0G-Y) in the center of the West edge of the matrix to an output LE cell (7,13,S1F-X) at the center of the East edge of the matrix. The evolvable region was from CLB 5,9 to 9,13 (in JBits row, col format). Evolution must also connect the input and output LEs to dedicated routing CLBs on the outside (of the evolvable region) neighbour.

Twenty evolutionary runs were done with a population size of 100 and a steady state genetic algorithm using tournament selection without replacement, for both morphogenetic and direct encoding approaches, with the only difference being in the decoding of the chromosome. The crossover rate was set at 80%, mutation at 2%, inversion (where a section of the chromosome is reversed) at 5%, and for the variable length chromosomes used with the morphogenetic approach, a base insert/delete rate of 0.1% was used with 50-50 chance of insertion or deletion. Each evolutionary run was continued until a solution with 100% fitness was found, or until 1000 generations had passed without an improvement in the maximum fitness attained.

For the morphogenesis approach, simulated transcription factors (TFs), including morphogens and cytoplasmic determinants, were disabled to determine the performance of morphogenesis based solely on hardware state. The initial population of chromosomes was pre-evolved for 25 generations for chromosome viability (see 5.1 for more details on this), and used a gene ceiling of 8 (which,

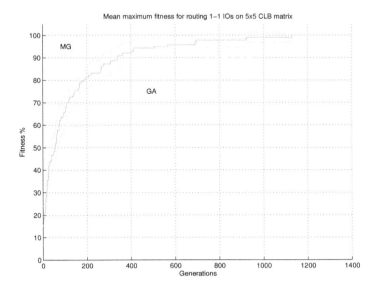

Figure A.1: Mean and SD maximum fitness for 5x5 CLB matrix: MG vs GA

however, doesn't prevent chromosomes having more than 8 genes).

Growth is done for a minimum of 30 growth steps, with fitness evaluated at each step, and growth continued if the maximum phenotype fitness for this genotype increased in the last 15 (minimum steps/2) growth steps, or if phenotype fitness is increasing. The genotype's fitness is given by the maximum phenotype fitness achieved during growth. The transcription rate was set to 4 bases per growth step, while the number of polymerase enzymes was set to be 30% of the number of genes (rounded to the closest integer), and the probability of a free polymerase binding to an activateable gene was set at 20%.

The results of these runs using the direct encoding and the morphogenetic approach are given in table A.1.

The direct encoding approach was able to find a 100% solution 13 out of the 20 runs, with the average number of generations required for successful runs being 531.08 generations with a standard deviation of 340.58. The morphogenetic approach was able to find a 100% solution every time, and took an average of 458.5 generations (with a standard deviation of 283.96), with the fittest phenotype occurring on average after 36.95 growth steps, and contained on average 9.9 genes on a chromosome length of 5690.35 bases.

The details of the fittest chromosomes evolved for solving the 5x5 routing problem are shown in table A.2. Note, however, that the used (by functional genes) length of chromosome, and percentage of chromosome used, are approximate values, due to the fact that the gene promoter count given by the statistics collection code (dchromstats.pl) may vary slightly from the used gene count as given by the morphogenesis system, if there are genes that encode no (complete) gene products.

## MG vs GA Multi Signal Routing on an 8x8 CLB Matrix

For the next set of experiments the size of the CLB matrix was increased to 8x8 (containing 256 cells), and the number of inputs and outputs to 4 each. The evolvable region was from CLB 4,8 to 11,15 (in JBits row, col format). Inputs are placed in the center of the West edge of the CLB matrix, 2 input locations per CLB (S0G-Y and S1F-X LEs at CLBs 7,8 and 8,8), while outputs are spread evenly across the East edge of the CLB matrix (at locations 4,15,S0G-Y; 6,15,S1F-X; 9,15,S0G-Y; 11,15,S1F-X). Here, also, the aim is to route signals, possibly inverted, from inputs to outputs, but in this case it was only necessary that each input and output is fully connected. In other words, the relationship between the different inputs and outputs is ignored, it is only important that all inputs are connected and one or more of these drive the outputs. These experiments increase the number of cells by 2.5 times, and the number of IO connections by 4, while requiring evolution to learn not just how to connect horizontally across the matrix, but also how to spread vertically from the middle outwards.

Again, twenty evolutionary runs were done with the same population sizes, evolutionary and morphogenetic parameters as the previous set of experiments. However, evolution was halted at success or when at least 5000 generations had been completed and 1500 generations passed with no increase in the maximum fitness attained within the population. The morphogenetic approach was again able to find a 100% solution on each run, and took an average of 1001.7 generations (standard deviation of 510.57), 49.95 growth steps, 5.65 genes and 3461.4 bases. However, the direct encoding approach was unable to find a 100% solution, with maximum fitness values reaching a mean of 86.64% with (standard deviation of 3.09%), and taking on average 4647.1 generations (standard deviation was 1756.9). The highest fitness achieved by the direct encoding approach was 93.75% which occurred at generation 9954. This run was continued up to 35,000 generations and reached a maximum of 96.875% at generation 16,302.

The results of these runs using the direct encoding and the morphogenetic approach are given in table A.1.

The details of the fittest chromosomes evolved for solving the 8x8 routing problem are shown in table A.2. As with the first set of experiments, note

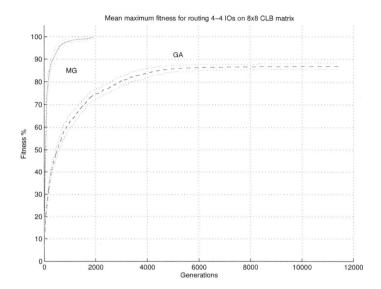

Figure A.2: Mean and SD maximum fitness for 8x8 CLB matrix: MG vs GA

Table A.1: Signal Routing Experiment Results

| Experiment Set | N | Max at Gen Mean | S.D. | Growth Steps Mean | S.D. | Max Fitness (%) Mean | S.D. |
|---|---|---|---|---|---|---|---|
| MG5x5 | 20 | 458.5 | 283.96 | 36.95 | 11.76 | 100.0 | 0.0 |
| MG8x8 | 20 | 1001.7 | 510.56 | 49.95 | 13.70 | 100.0 | 0.0 |
| GA5x5 | 20 | 410.80 | 323.53 | - | - | 93.00 | 9.79 |
| GA5x5 100% only | 13 | 531.08 | 340.58 | - | - | 100.00 | 0.0 |
| GA8x8 | 20 | 4023.2 | 694.1 | - | - | 85.51 | 3.06 |
| GA8x8 > 5000 | 20 | 4647.1 | 1756.9 | - | - | 86.64 | 3.09 |

that the used length of chromosome, and percentage of chromosome used, are approximate values. "Enh." and "Rep." are shorthand for enhancer and repressor (length or bind sites per gene), respectively.

Figures A.1 and A.2 show the mean maximum fitness (and standard deviation as the dotted lines) over all runs for both approaches on the two experiments. Note that in the latter generations, particularly in Figure A.2, most experiments have stagnated (fitness hasn't increased for 1500 generations) and

**203**

Table A.2: Details of Fittest Evolved Chromosomes for Morphogenesis on 5x5 and 8x8 CLB Matrix

| Experiment Set | Mean Length | | | | | | Gene Count | Sites/Gene | |
|---|---|---|---|---|---|---|---|---|---|
| | Chrom | Used | Used% | Gene | Enh. | Rep. | | Enh. | Rep. |
| 5x5 Mean | 5690.35 | 969.24 | 17.9 | 66.76 | 21.83 | 21.10 | 9.90 | 0.734 | 0.558 |
| 5x5 S.D. | 4965.68 | 780.69 | 4.0 | 10.02 | 6.90 | 8.57 | 7.91 | 0.217 | 0.213 |
| 8x8 Mean | 3461.40 | 601.90 | 19.3 | 71.32 | 21.83 | 20.14 | 5.65 | 0.917 | 0.664 |
| 8x8 S.D. | 1843.21 | 292.93 | 7.5 | 13.54 | 6.90 | 10.05 | 2.94 | 0.609 | 0.314 |

so been stopped, hence the highest fitness achieved in those runs is used for calculating the mean and standard deviation across all runs at that generation. Although, it is possible that some fitness increase could occur in stagnant populations if those runs were continued, in practice this has never occurred, and due to time constraints, this approach has been taken as a reasonable estimation.

## Morphogenesis vs Direct Encoding Results Summary

In both sets of experiments the morphogenesis approach was able to find a 100% solution every time. Morphogenesis took an average of 458.5 generations to solve the first experiments, with a standard deviation (SD) of 283.96 generations, while in the second set of experiments an average of 1001.7 generations was required (SD=510.56).

In the first set of experiments the direct encoding approach was able to find a 100% solution in 13 out of the 20 runs, with the average number of generations required for successful runs being 531.08 generations (SD=340.58). For the second set of experiments, this approach was unable to find any 100% solution, with maximum fitness values reaching a mean of 86.64% (SD=3.09%), and taking on average 4647.1 generations (SD=1756.9). The best run, which reached 93.75% at generation 9954, was continued up to 35,000 generations, reaching a maximum of 96.875% at generation 16302.

Figure A.3 shows the mean maximum fitness over all runs for both approaches on the two experiments (up to generation 5000). From this it is evident that the morphogenetic approach (denoted by MG) not only outperforms the direct encoding approach (denoted by GA), but also scales very well, with the more complex problem (MG8x8) keeping pace with the simpler problem (MG5x5) up until the last few fitness percentage points, which corresponds to connecting all 4 outputs, at which stage a few runs lag behind.

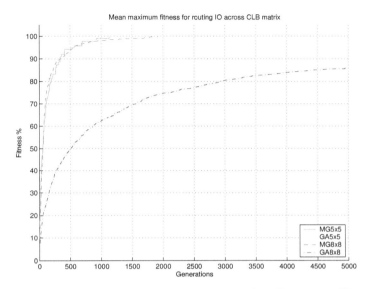

Figure A.3: Mean max fitness for routing IO across 5x5 and 8x8 CLB matrix: MG vs GA

## A.2  Comparison of 5x5 and 8x8 Morphogenesis Runs

In both sets of experiments the morphogenesis approach was able to find a 100% solution every time. While there was an approximately 2.5-fold increase in the number of cells from 100 (5x5 CLBs with 4 cells per CLB) to 256 (8x8 CLBs with 4 cells per CLB), and a 4-fold increase in the number of inputs and outputs (from 2 to 8), the number of generations required to solve the problem only doubled. Likewise, the number of growth steps required only had a modest (30%) increase.

In the more complex problem, the average chromosome length decreased to 60% of that required for the simpler problem, and the number of genes also decreased by a similar amount. These changes in chromosome structure, between the two sets of experiments, seemed to reflect an increasing refinement in the encoded morphogenetic process as the problem complexity increases.

Table A.3 provides the details of the 100% solution chromosomes for 5x5 and 8x8 runs. Note that, although TFs were provided in the genetic code, TF binding (to regulate genes) wasn't used in these runs.

Table A.3: Comparison of Morphogenetic Runs

| Feature | 5x5 Mean / 8x8 Mean / $\frac{8x8}{5x5}$ Mean | 5x5 Std Dev. / 8x8 Std Dev. | 5x5 Notes / 8x8 Notes |
|---|---|---|---|
| Soln at Generation | 458.5 / 1001.7 / 218.47% | 283.96 / 510.56 | |
| Soln at Growth Iter. | 36.95 / 49.95 / 135.18% | 11.76 / 13.70 | |
| Gene Count | 9.90 / 5.65 / 57.07% | 7.91 / 2.94 | No. RNApoly=2.97 (SD=2.37) / No. RNApoly=1.70 (SD=0.88) |
| Genes Transcribed (Approx.) | 5.06 / 3.72 / 73.52% | | 32.89 gene prods = 25.22 used res / 25.97 gene prods = 19.91 used res / (prods=products; res=resources) |
| Enhancer Bind Sites Per Gene | 0.734 / 0.917 / 124.93% | 0.217 / 0.609 | |
| Repressor Bind Sites Per Gene | 0.558 / 0.664 / 119.10% | 0.213 / 0.314 | |
| Mean Gene Length | 66.76 / 71.32 / 106.83% | 10.02 / 13.54 | 6.50 (SD=1.05); 16.69 (SD=2.5) / 6.98 (SD=1.43); 17.83 (SD=3.4) / (gene products; transcription steps) |
| Mean Enhancer Length | 21.83 / 21.83 / 100.00% | 6.90 / 6.90 | 4.24 (SD=1.34) used resources / 4.51 (SD=1.42) used resources |
| Mean Repressor Length | 21.10 / 20.14 / 95.45% | 8.57 / 10.05 | 4.10 (SD=1.66) used resources / 4.16 (SD=2.07) used resources |
| Chromosome Length | 5690.35 / 3461.4 / 60.83% | 4965.68 / 1843.21 | |
| Used Chromosome Length | 969.24 / 601.90 / 62.10% | 780.69 / 292.93 | |
| Total Gene Coding Length | 673.87 / 391.48 / 58.09% | 570.83 / 197.79 | 11.84% of chrom / 11.31% of chrom / 95.5% |
| Total Enhancer Bind Length | 146.24 / 111.30 / 76.11% | 112.40 / 77.00 | 02.57% of chrom / 03.22% of chrom / 125.3% |
| Total Repressor Bind Length | 109.53 / 76.52 / 69.87% | 91.40 / 53.50 | 01.92% of chrom / 02.21% of chrom / 115.1% |
| Used % of Chromosome | 17.93 / 19.27 / 107.49% | 4.04 / 7.50 | junk = 82.07% / junk = 80.63% |

Note that the used length of chromosome (used by functional genes), and hence also the percentage used, are approximate values, due to the fact that the gene promoter count (indicating genes that are transcribable) given by *dchromstats.pl* (`genep`) may vary slightly from the used gene count as given by the morphogenesis system (`Gcnt`), as some genes may contain no gene products after the translated chromosome has been converted by *resattrset2bitsval.pl*.

## Calculation of Gene and Chromosome Feature Estimates

Used length was produced with the following Matlab command (and rounded to the closest integer).

```
usefl=(glenm + (elenm .* ebdvg) + (rlenm .* rbdvg) + plen) .* Gcnt
```

where,

**Gcnt** = gene count,

**plen** = promoter length = 4 (TATA),

**glenm** = mean gene coding region length,

**elenm** = mean length of enhancer bind sites,

**rlenm** = mean length of repressor bind sites,

**ebdvg** = mean number of enhancer bindsites / `genep`,

**rbdvg** = mean number of repressor bindsites / `genep`

The used percentage was calculated as (`usefl ./ len`) `* 100` and then rounded to the nearest integer. Total gene coding length, total enhacer bind site length, and total repressor bind site length were calculated with the Matlab commands

**totgcl** = $glenm. * Gcnt$

**totebl** = $(elenm. * ebdvg). * Gcnt$

**totrbl** = $(rlenm. * rbdvg). * Gcnt$

For all runs the transcription parameters were:

- transcription rate $T = 4$

- polymerase to gene ratio $P = 0.3$

- polymerase bind activation threshold $A = 0.8$

Table A.4: Resource Codon Allocations in Genetic Code

| Resource Type | Allocated Entries | Codons to Encode | Codons × Entries |
|---|---|---|---|
| SliceIn | 18 | 3 | 54 |
| LUTBitFN | 4 | 5 | 20 |
| SliceToOut | 8 | 2 | 16 |
| OutToSingle | 16 | 2 | 32 |
| Used Resources | 46 | | 122 |
| LUTIncrFN | 8 | 4 | 32 |
| SliceRAM | 2 | 2 | 4 |
| FPGA Resources | 56 | | 158 |
| TF, Local | 2 | 2+ (8) | 16 |
| TF, Morphogen | 2 | 4+ (8) | 16 |
| Coded Resources | 60 | | 190 |

The following was used to estimate the number of encoded resources. For `glenm` note that 6 is subtracted from the gene length for the start + stop codons, while for `elenm` and `rlenm` total bases for encoding resources is given as:

$$B = \frac{2}{3}(\text{sampling}) \cdot 3(\text{interleaving}) \cdot length = 2 \cdot length \qquad (A.1)$$

The number of codons required to encode the different resources is given in Table A.4. Note that TFs are encoded in a variable number of codons, so an approximation of 8 was chosen, based on an examination of decoded chromosomes.

From this the average length in bases required to encode used FPGA resources can be estimated as the sum of the number of codons required (122) to encode every distinct used FPGA resources (46). This gives an average

$$RU = \frac{122}{46} = 2.65 \quad \text{codons} = 7.96 \quad \text{bases} \qquad (A.2)$$

The average length required to encode resources in general is calculated in the same manner

$$R = \frac{190}{60} = 3.17 \quad \text{codons} = 9.5 \quad \text{bases} \qquad (A.3)$$

and the average number of bases required to encode FPGA resources (excluding TFs, but including unused resources) is

$$RF = \frac{158}{56} = 2.82 \quad \text{codons} = 8.46 \quad \text{bases} \qquad (A.4)$$

With a transcription rate ($T = 4$) bases per growth step the average number of growth steps required to transcribe a gene product can be calculated as

$$TR = \frac{R}{T} = \frac{9.5}{4} = 2.375 \qquad (A.5)$$

The percentage of actually used (FPGA only) resources

$$UR = \frac{46}{60} = 76.67\% \qquad (A.6)$$

while the ratio of used FPGA resources (46) to all FPGA resources (56) is

$$FR = 46/56 = 82.14\% \qquad (A.7)$$

With the bind activation threshold $A = 0.8$, the probability of a polymerase enzyme binding at a given time step is

$$PB = 1.0 - 0.8 = 0.2 = 20\% \qquad (A.8)$$

then on average the number of growth steps before the polymerase enzyme binds for transcription (again)

$$TA = \frac{1}{PB} = \frac{1}{0.2} = 5 \qquad (A.9)$$

Using this, some estimates can then be made on the evolved chromosomes that attained 100% fitness, and their expected behaviour during a morphogenetic run. This is done for both the 5x5 and 8x8 experiment sets. Note that in the following, results are rounded to 2 decimal places.

For 5x5 runs, the mean gene length is

$$GL = 66.76 \quad \text{bases} \qquad (A.10)$$

so the average number of growth steps required to transcribe a gene is given by

$$TG = \frac{GL}{t} = \frac{66.76}{4} = 16.69 \qquad (A.11)$$

With a polymerase to gene ratio of $P = 0.3$, the number of polymerase enzymes in the cell is given as

$$PN = GN \cdot P = 9.90 \times 0.3 = 2.97 \qquad (A.12)$$

Then, given that the average number growth steps per run

$$S = 36.95 \qquad (A.13)$$

and the average number of growth steps before a polymerase enzyme binds again is given by Eqn. A.9, the average number of genes transcribed in a morphogenesis run can be estimated as

$$
\begin{aligned}
TTG &= \frac{S}{TA+TG} \cdot PN \\
&= \frac{36.95}{5+16.69} \times 2.97 = 5.06
\end{aligned}
\tag{A.14}
$$

The average number of resources encoded on a gene is equal to the gene coding length minus the start and stop codons and then divided by the average length of an encoded resource. Hence,

$$
RG = \frac{GL-6}{R} = 67.76 - 69.5 = 6.50
\tag{A.15}
$$

Then the number of resources configured during a run can be estimated as the number of resources encoded on a gene multiplied by the number of genes expressed. This is given as

$$
C = RG \cdot TTG = 5.06 \times 6.5 = 32.89
\tag{A.16}
$$

Not all of these resources are used, so to find the actual number of configured FPGA resources $C$ is scaled by the percentage of used resources ($UR$) as given in Table A.4. This gives the number of FPGA resources configured in a single cell during a morphogenesis run as

$$
CU = C \cdot UR = 32.89 \times 0.7767 = 25.22
\tag{A.17}
$$

The mean length of enhancer bind sites was determined to be

$$
El = 21.83
\tag{A.18}
$$

while that of repressor bind sites was

$$
Rl = 21.10
\tag{A.19}
$$

In estimating the number of resources encoded on an average enhancer or repressor bind site (TFs are ignored as they are treated differently by the morphogenesis system), remembering that the total length of binding bases on a bindsite is $length \cdot 2$ as elaborated in Equation A.1. Using these gives us the average number of resources encoded on an enhancer bind site

$$
RE = \frac{El \cdot 2}{RF} = 21.83 \times 28.46 = 5.16
\tag{A.20}
$$

and on the repressor bind site

$$RR = \frac{Rl \cdot 2}{RF} = 21.10 \times 28.46 = 4.99 \qquad (A.21)$$

These are then scaled according to the ratio of used FPGA resources to all FPGA resources to determine the number of used FPGA resources that bind (and thus contribute to regulating any given gene).

$$EB = RE \cdot FR = 5.16 \times 0.82 = 4.24 \qquad (A.22)$$

and

$$RB = RR \cdot FR = 4.99 \times 0.82 = 4.1 \qquad (A.23)$$

For 8x8 runs the same procedure is repeated to give the estimate the values as follows:

Average gene length is given as

$$GL = 71.32 \quad \text{bases} \qquad (A.24)$$

the average number of growth steps required to transcribe a gene is then

$$TG = \frac{GL}{T} = \frac{71.33}{4} = 17.83 \qquad (A.25)$$

the number of polymerase enzymes in the cell is given as

$$PN = GN \cdot P = 5.65 \times 0.3 = 1.70 \qquad (A.26)$$

With an average number growth steps per run

$$S = 49.95 \qquad (A.27)$$

the average number of genes transcribed in a morphogenesis run is

$$\begin{aligned} TTG &= \frac{S}{TA + TG} \cdot PN \\ &= \frac{49.95}{5 + 17.83} \times 1.70 = 3.72 \end{aligned} \qquad (A.28)$$

The average number of resources encoded on a gene is

$$RG = \frac{GL - 6}{R} = 71.32 - 69.5 = 6.98 \qquad (A.29)$$

so the number of resources configured during a run will be

$$C = RG \cdot TTG = 6.98 \times 3.72 = 25.97 \qquad (A.30)$$

while the number of FPGA resources configured in a single cell during a morphogenesis run is

$$CU = C \cdot UR = 25.97 \times 0.7767 = 19.91 \tag{A.31}$$

The mean length of enhancer bind sites was determined to be

$$El = 21.83 \tag{A.32}$$

while that of repressor bind sites was

$$Rl = 20.14 \tag{A.33}$$

Using these, combined with Equations A.1 and A.4 gives us the average number of resources encoded on an enhancer bind site

$$RE = \frac{El \cdot 2}{RF} = 21.83 \times 28.46 = 5.49 \tag{A.34}$$

and on a repressor bind site

$$RR = \frac{Rl \cdot 2}{RF} = 20.14 \times 28.46 = 5.06 \tag{A.35}$$

Based on this and Equation A.7, the number of used FPGA resources that bind on any given gene is

$$EB = RE \cdot FR = 5.49 \times 0.82 = 4.51 \tag{A.36}$$

and

$$RB = RR \cdot FR = 5.06 \times 0.82 = 4.16 \tag{A.37}$$

# A.3    Simulated Transcription Factor Experiments

## A.3.1    MG vs TF Initial Test

The TF parameters used for the first set of experiments were as follows:

- Morphogen propagation delay $= 1$

- Morphogen TF initial time to live $= 8$

- Local TF initial time to live $= 8$

- Unbound TF age rate $= 3$

- Bound TF age rate $= 5$

Table A.5: Results of Morphogenesis with TFs for Signal Routing

| Experiment | N | Max at Gen | | Growth Steps | | Gene Count | | Max Fitness (%) | |
| Set | | Mean | S.D. | Mean | S.D. | Mean | S.D. | Mean | S.D. |
|---|---|---|---|---|---|---|---|---|---|
| TF5x5 | 10 | 485.5 | 282.6 | 35.4 | 7.03 | 8.0 | 4.52 | 100.0 | 0.0 |
| TF8x8 | 10 | 959.3 | 596.9 | 51.4 | 6.93 | 4.5 | 1.43 | 99.7 | 0.94 |
| TF8x8 2K | 10 | 1145.1 | 717.2 | 53.2 | 9.53 | 4.6 | 1.35 | 100.0 | 0.0 |

According to these parameters, morphogens will spread at a rate of 1 CLB per growth step, however, there was a bug in the MGEHW cell implementation, which meant that TFs were aged as soon as they were released in the cell and prior to testing for binding. The effect of this was that the initial time to live was effectively equal to 5, meaning that an unbinding TF will live for another 2 growth steps after being released in the cell (as a gene product or through morphogen spread) if it doesn't bind to any regulatory region, or another 1 growth step if it does bind.

Table A.5 shows the results of running the morphogenetic EHW system on the 5x5 and 8x8 experiment sets. 10 runs were done for each set. The same morphogenetic and evolutionary parameters were used as for the previous experiments that relied solely on FPGA state.

Note that one of the 8x8 experiment set runs stagnated after 1500 generations, giving a 90% success rate, but when stagnation was increased to 2000 generations, a 100% solution was found. The mean and standard deviation shown in Table A.5 with the increased stagnation parameter are indicated with a "2k" suffix.

This first set of experiments were run before any knowledge of reasonable values for these parameters could be ascertained, and furthermore, were chosen when an earlier set of morphogenetic parameters were being used, in which there were a higher number of polymerase enzymes in the cell and a greater probability of binding, which would have resulted in TFs' life spans not needing to be very long to have an affect on gene regulation.

With the knowledge gained from these experiments, a further set of experiments with reasonable TF parameters was run to determine the usefulness of simulated molecules in the morphogenetic EHW system.

## A.3.2    MG vs TF Test Rerun

For TFs to have any effect, they need to live at least as long as would be required to allow polymerase to bind. With around 5 growth steps required for a given free polymerase to bind to a given gene, and approximately 4 growth steps to transcribe a gene on average (from Table A.3), then the worst case may be around 9 growth steps from TF binding to when polymerase is activated for that gene, if all polymerase are in use.

Given approximately 1.5 TFs per gene (TF resources given in Table A.4 and average gene information in Table A.3) and 4 growth steps to transcribe a gene, with a 30% polymerase to gene ratio and an average of say 8 genes, then there would be approximately 2 genes being transcribed at any time ($8/3 = 2.67$ which is rounded down to 2 polymerase), giving $2 \times 1.5/4$ TFs per growth step, which is approaching one TF produced per growth step.

For unbound TFs, which would either be unbindable morphogens or duplicates of bound TFs, then a shorter life span would be more suitable. Possibly 1–2 growth steps (to avoid buildup of molecules and associated computational burden). Given that unbound TFs have a short life span, a longer life span, of 4–6 growth steps, for bound TFs would be reasonable to ensure that they have a chance of triggering the expression of genes.

From this, an unbound TF lifespan of 2 growth steps, and a bound TF lifespan of 5 should be suitable for discerning the effect of adding TFs to the morphogenetic EHW system. This gives the following TF parameters, which were used for these experiments:

- Morphogen propagation delay = 1

- Morphogen TF initial time to live = 10

- Local TF initial time to live = 10

- Unbound TF age rate = 5

- Bound TF age rate = 2

According to these parameters, morphogens will spread at a rate of 1 CLB per growth step, and a TF will live for another 2 growth steps after being released in the cell (as a gene product or through morphogen spread) if it doesn't bind to any regulatory region, or another 5 growth steps if it does bind.

### MG vs TF Test Run 2 Results

Table A.6 shows the results of running the morphogenetic EHW system on the 5x5 and 8x8 experiment sets. 10 runs were done for each set. The same morphogenetic and evolutionary parameters were used as for the previous experiments that relied solely on FPGA state.

Note that two runs of the 8x8 experiment set stagnated after 1500 generations, giving a 80% success rate, however, unlike the first set of TF experiments, when the stagnation parameter was increased to 2000 generations, these runs were still unable to make any progress towards finding a 100% solution.

Table A.6: Results of Morphogenesis with Longer Life TFs for Signal Routing

| Experiment Set | N | Max at Gen | | Growth Steps | | Gene Count | | Max Fitness (%) | |
|---|---|---|---|---|---|---|---|---|---|
| | | Mean | S.D. | Mean | S.D. | Mean | S.D. | Mean | S.D. |
| TF5x5 | 10 | 402.3 | 222.1 | 39.3 | 6.8 | 7.2 | 3.39 | 100.0 | 0.0 |
| TF8x8 | 10 | 1063.1 | 329.5 | 53.6 | 14.3 | 4.0 | 2.36 | 99.375 | 1.32 |

Table A.7: Comparison of Morphogenetic Runs with and without Simulated TFs

| Experiment Set | N | Max Fitness | | Max at Gen | | Growth Steps | | Genes | |
|---|---|---|---|---|---|---|---|---|---|
| | | Mean | S.D. | Mean | S.D. | Mean | S.D. | Mean | S.D. |
| MG5x5 | 20 | 100 | 0.0 | 458.5 | 284.0 | 37.0 | 11.8 | 9.9 | 7.9 |
| TF5x5 | 10 | 100 | 0.0 | 408.5 | 282.6 | 35.4 | 7.0 | 8.0 | 4.5 |
| TF2r5x5 | 10 | 100 | 0.0 | 402.3 | 222.1 | 39.3 | 6.8 | 7.2 | 3.4 |
| MG8x8 | 20 | 100 | 0.0 | 1001.7 | 510.6 | 50.0 | 13.7 | 5.7 | 2.9 |
| TF8x8 | 10 | 99.7 | 1.0 | 959.3 | 596.9 | 51.4 | 6.9 | 4.5 | 1.4 |
| (cont) | 10 | 100 | 0.0 | 1145.1 | 717.2 | 53.2 | 9.5 | 4.6 | 1.3 |
| TF2r8x8 | 10 | 99.4 | 1.3 | 1063.1 | 329.5 | 53.6 | 14.3 | 4.0 | 2.4 |
| (100%) | 8 | 100 | 0.0 | 1143.3 | 318.7 | 53.0 | 15.0 | 4.1 | 2.6 |

## A.3.3   MG vs TF Results and Discussion

Table A.7 compares the results of running these tests with TFs ("TF2r" is used to denote the second TF run with longer TF lifespans) against the previous runs, in which TFs weren't used; and Figures A.4 and A.5 show the mean maximum fitness over all runs for the different approaches on the 5x5 and 8x8 experiment sets.

From the results shown in Figure A.4, it appeared that the addition of TFs was slightly beneficial on the smaller CLB matrix, especially when TFs had a longer lifespan. However, on the larger problem, the addition of TFs was detrimental in two ways. Firstly, it appeared that the stronger the influence of TFs on gene regulation, through increased TF lifespan, the higher the likelihood of failure to find a 100% solution (interestingly, stagnation always occurred at the 96.875% mark, at which point one output is not connected), and secondly the average performance of only successful TF runs was considerably worse than if TFs weren't used (around 15% more generations were required to find a 100% solution).

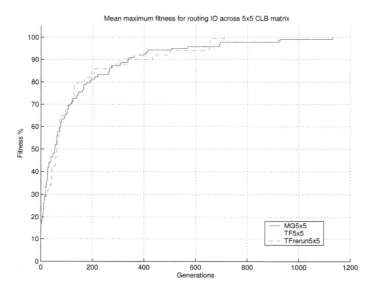

Figure A.4: Mean maximum fitness for 5x5 CLB matrix MG and TF runs

## A.4    Speed Tests

To compare the computational performance of the different approaches, another set of runs was conducted, one for each approach (GA, MG, TF) on both experiments. Runs were done for 100 generations on a 1.5 GHz Pentium 4, with 256KB cache, 400 MHz system bus, 512MB RAM and running Windows 2000 Professional. The same population size (100) and parameters used in the previous experiments were used here. The results are summarised in Table A.8. Note that GA setup took 0 minutes, while MG and TF setup included 25 generations of pre-evolution for chromosome viability (see 5.1, and took 2-3 minutes. The mean gene counts over generations 1–100 was calculated as the sum of gene counts in all chromosomes in the population for each generation, divided by the number of generations (100). Also note that the TF parameters used were those of the first set of TF experiments, discussed in Section 7.3.

From this it can be seen that while the direct encoding approach is more than twice as fast per generation for the smaller problem, and 3–4 times as fast for the larger, this is offset by the number of generations required for the direct encoding approach to solve the given problem. Thus, for the 5x5 problem the direct encoding approach was twice as fast as the morphogenetic approach,

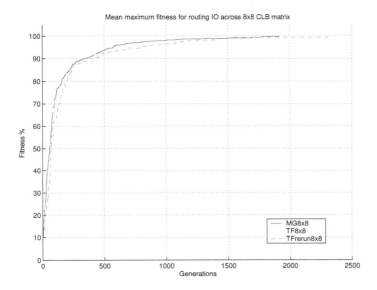

Figure A.5: Mean maximum fitness for 8x8 CLB matrix MG and TF runs

Table A.8: Speed Tests For Routing Experiments Over 100 Generations

| Experiment Set | Time to Complete Gen | | | Max Gen | Time to Max Gen | | Avg Gene Count |
|---|---|---|---|---|---|---|---|
| | 0 (min) | 100 (min) | 1000 (h:m) | | min | hr:mn | |
| GA5x5 | 02 | 53 | 8:50 | 531.1 | 282 | 04:42 | - |
| MG5x5 | 06 | 120 | 20:00 | 458.5 | 550 | 09:50 | 6.78 |
| TF5x5 | 07 | 149 | 24:50 | 408.5 | 609 | 10:09 | 8.42 |
| GA8x8 | 02 | 58 | 9:40 | 4647.1 | 2695 | 44:55 | - |
| | | | | 5000 | 2900 | 48:20 | - |
| MG8x8 | 08 | 188 | 31:20 | 1001.7 | 1883 | 31:23 | 7.35 |
| TF8x8 | 13 | 240 | 40:00 | 1145.1 | 2748 | 45:48 | 5.56 |

but on the 8x8 problem it took 1.5 times as long to reach the generation at which maximum fitness was attained. Further, the direct encoding approach was unable to solve the 8x8 problem in any run, even when run for 36,000 generations (2 weeks according to the rates indicated in the table).

Hence the morphogenesis system shows its value once problem complexity is increased. Although there is a modest increase in computational cost (25%)

when the number of cells was more than doubled (from 100 to 256 cells), it should be noted that the morphogenesis process here was implemented in software on a sequential processor. If the morphogenesis process itself is implemented in hardware, then increasing the number of cells should not incur any additional cost in processing time (although, of course there is a cost in terms of hardware "real estate"). Furthermore, the execution time of the morphogenesis process would be decreased significantly, due to the parallelism offered by hardware (for a start), and would be competitive with the direct encoding, on a per-generation metric.

In all cases, the morphogenesis approach that incorporated simulated TFs, had the highest computational cost per-generation and per-average evolutionary run. It is possible, however, that with a better choice of parameters for TFs, then an improvement in the number of generations taken to solve a given problem will offset the higher per-generation cost.

# A.5    Sensitivity Analysis

A sensitivity analysis was performed on both evolutionary and morphogenetic parameters (but not on TF parameters, as simulated TFs were not used here) for the 5x5 problem. All runs used as their starting population the generation 0 population (i.e. the initial population, which in the case of morphogenesis has already completed 25 generations of pre-evolution to seed the population with "viable" chromosomes) that was generated by the closest run to the mean, in terms of the number of generations required to find a 100% solution (these are referred to in the tables as the average run), and were run to completion or stagnation (1000 generations without an improvement in maximum fitness).

## A.5.1    Sensitivity of Evolutionary Parameters

The results of the first set of runs in which evolutionary parameters were varied for both the direct encoding and morphogenetic approaches is shown in Tables A.9 and A.10. Entries with a '*' beside them indicate that stagnation occurred at that generation, and so evolution was halted (after completing that generation). For all runs that deviated from the mean by at least 1 S.D., the likelihood of that result occurring within the original experiments, is shown as a percentage.

From this it is evident that very high mutation rates (8%) are deleterious, as expected. It also appears, from these runs, that the morphogenetic approach prefers low rates of mutation and high rates of crossover, while for the direct encoding approach crossover and mutation rates seem to have preferred ranges around the rates used for the experiments in this chapter, with mutation rates working well within the 2-4% range and also below the 1% range at 0.5%, but

Table A.9: Sensitivity Analysis on Evolutionary Parameters

| | x=80,m=2,i=5,id=0.1 | | | Crossover | | Mutation | | | | Inversion | |
|---|---|---|---|---|---|---|---|---|---|---|---|
| | Mean | SD | Avg Run | 60 | 100 | 0.5 | 01 | 04 | 08 | 0 | 10 |
| GA | 531 | 341 | 572 | 1173* | 1308* | 711 | 1171* | 483 | 1899 | 882 | 451 |
| - mean | | | +41 | +642* | +777* | +180 | +640* | -48 | +1368 | +351 | -80 |
| z-value | | | +0.1 | > 1.9 | > 2.3 | +0.5 | > 1.9 | -0.1 | 4.0 | +1.0 | -0.2 |
| likelihood | | | | < 5% | < 2% | | < 5% | | ≈ 0% | 32% | |
| MG | 459 | 283 | 414 | 1018 | 260 | 256 | 116 | 838 | 1186* | 898 | 217 |
| - mean | | | -45 | +559 | -199 | -203 | -343 | +379 | +727* | +439 | -242 |
| z-value | | | -0.2 | +2.0 | -0.7 | -0.7 | -1.2 | +1.3 | > 2.6 | +1.6 | -0.9 |
| likelihood | | | | 4% | | | 24% | 20% | < 1% | 10% | |

Table A.10: Sensitivity Analysis on Evolutionary Parameters (cont.)

| | x=80,m=2,i=5,id=0.1 | | | ins/del | |
|---|---|---|---|---|---|
| | Mean | SD | Avg Run | 00 | 01 |
| MG | 459 | 283 | 414 | 528 | 920 |
| - mean | | | -45 | +69 | +461 |
| z-value | | | -0.2 | +0.2 | +1.6 |
| likelihood | | | | | 10% |

with the intervening 1% range demonstrating poor performance. The use of the inversion operation, at the standard rate and higher, seems to be beneficial especially for the MG approach, and removal of this operator appeared to have some detriment to performance. Higher rates of the base insertion/deletion operator used in the MG approach decreased performance, while removal of this operator had little effect.

## A.5.2    Sensitivity of Morphogenetic Parameters

The results of the second set of runs in which morphogenetic parameters were varied for the morphogenetic approaches is shown in Table A.11. The number of generations required to find an 80% solution are also shown. Only three parameter sets were used, these being the "standard" set that was used in the experiments presented in this chapter; a "hot" (more chaotic) set in which there are a larger number of polymerase enzymes per gene ($P$), an increased probability of a free polymerase binding to a free gene by decreasing the polymerase bind activation threshold ($A$) for initiating transcription (pt=0.6 gives 40% probability of binding), and to offset this a slower transcription rate ($T$); and lastly a "cool" set with fewer polymerase, decreased probability of binding (10% chance) by increasing the binding threshold, and an increased transcrip-

Table A.11: Sensitivity Analysis on Evolutionary Parameters

| T=4,P=0.3,A=0.8 | | | T=2,P=0.5,A=0.6 | T=6,P=0.1,A=0.9 |
|---|---|---|---|---|
| Mean | S.D. | Avg Run | Hot Run | Cool Run |
| 459 | 283 | 414 | 1059* | 1631 |
| | - mean | - 45 | +600* | +1172 |
| | z-value | -0.2 | > 2.12 | +4.14 |
| | likelihood | | < 3% | ≈ 0% |
| 80% Mark | | 166 | 59 | 1225 |

Table A.12: Sensitivity Analysis for Anti Gene Bloat

| | AGB | | No AGB | |
|---|---|---|---|---|
| | Mean | SD | Run | Variance |
| Gens | 458.5 | 283.96 | 590 | +0.46 |
| Growth Steps | 36.95 | 11.76 | 56 | +1.6 |
| Gene Count | 9.90 | 7.91 | 3 | -0.87 |

tion rate to offset this.

From this it can be seen that increases in polymerase to gene ratio and decreases in polymerase binding threshold have the effect of "heating up" the system, in spite of the moderating affect of the slower transcription rate. This caused the system to become more chaotic, performing more of a brute force attack, which allows it to find good solutions early on (the 80% mark), but makes it harder to a more refined search, as required to find a 100% solution (as shown by the run stagnating).

In contrast, the run with a decreased polymerase to gene ratio, and increased binding threshold, with an increased transcription rate to moderate the affects of these, performed a slow but steady search that prevented stagnation and eventually brought it to a 100% solution, admittedly requiring close to 3 times as many generations as an average run would take.

## A.5.3    Sensitivity to Anti-Gene-Bloat

To test the effect of the anti-gene-bloat mechanism, another run of the MG system was done with the same starting population as before, and with the standard MG parameters. The results of this run are given in Table A.12.

This run took 590 generations to complete, being within 0.46 SD of the mean, with the 100% solution taking 56 growth steps, and having 3 genes, which surprisingly was 6.9 below the mean, but statistically speaking this is only -0.87 SD. During this run the average gene count increased from 7.99 at generation 0 to 10.94 at generation 590. At generation 410 the gene counts of chromosomes in the population ranged between 2 and 38, with a mean of 13.33 and SD of 8.31. In comparison, at generation 410, the average MG run, with a soft gene ceiling of 8, had chromosomes with between 1 and 7 genes, with a mean of 3.91 and a SD of 1.63.

So, while there was some small increase in gene counts during the morphogenetic run, it wasn't much of a problem here, although this run was relatively short. Also, due to the low polymerase to gene ratio (0.3) and high bind threshold (0.8), this probably prevents chromosomes with high gene counts having much advantage, as few genes are able to be activated.

## A.6   Summary of Results

In the experiments run to date, the morphogenetic approach tends to maintain a more diverse population, in terms of fitness spread, than that of the direct encoding. The direct encoding approach also tends to make frequent small fitness increments, which decrease in frequency but maintain magnitude during an evolutionary run. The morphogenetic approach, however, tends to make less frequent, but more pronounced fitness jumps, indicative of punctuated equilibrium, whereby neutral mutations collect, until able to make an evolutionary leap towards a fitter phenotype. Morphogenetic chromosomes also appear to respond to increases in problem complexity through increasing refinement of the encoded genetic information.

As demonstrated by the experiments conducted here, the morphogenetic approach is consistently able to outperform a traditional direct encoding approach to evolvable hardware. On average the morphogenetic approach finds a fitter solution in less generations, and shows less likelihood of getting caught in local maxima.

The ability of the morphogenetic approach to scale with increased size and problem complexity was demonstrated in the second set of experiments, in which the number of cells was increased by 2.56 times, and IO connections by 4 times, while requiring evolution to learn how to spread and connect in several directions. Despite this increase in size and complexity, the morphogenesis approach only required, on average, 2.18 times as many generations (1.35 times as many growth steps and close to half as many genes) to solve this problem, while no direct encoding run was able to solve it demonstrating that the morphogenetic approach not only works, but scales well.

While the morphogenesis approach takes longer on simpler problems than a traditional EHW approach (typically half again as long on the first set of experiments), on more complex problems its faster evolutionary progress outweighs its higher per-chromosome evaluation cost, while the direct encoding approach becomes increasing computationally unviable.

## A.6    Summary of Results

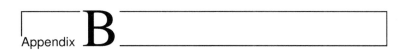

# Experiment Complexity Calculations

In Section 9.1 of Chapter 9, measures of circuit complexity were given for calculating configuration state space, fitness feedback and problem difficulty.

In this appendix, details of the calculations, and the architectural and problem specific details required to arrive at these, are provided for each set of experiments.

Prior to this, however, it is worth noting the following. As the state spaces are generally quite large and probabilities quite small, calculations would tend to generate numbers that would need to be presented in some form of logarithmic notation, such as scientific notation, for practical and informational purposes. However, as FPGAs are digital devices, this means that configuration spaces are basically binary in nature, as are feedback spaces. For this reason, all calculation results are presented as base-2 logarithms (generally rounded to 3 decimal places for presentation, though higher precision is used in the actual calculations).

## B.1 Routing Experiment Calculations

### B.1.1 Calculating State Space

All of the routing experiments use a slimmed down set of FPGA resources, in which each cell, mapped to a logic element, has:

- one LUT input with 4 (S1G-Y) or 5 available settings: 3 (for S1G1) or 4 single lines and OFF;

- 4 LUT configurations: pass, invert, always 0, always 1;

- 2 output buses with 2 states: logic element's registered output (e.g S1_YQ) or OFF

- 3 (for S1G-Y) or 4 output lines, each with with 2 states: ON or OFF.

Table B.1: Routing Experiment State Spaces

| CLB Matrix | CLBs | S ($log_2$) |
|---|---|---|
| 5x5 | 25 | 999.145 |
| 8x8 | 64 | 2557.810 |
| 9x9 | 81 | 3237.229 |
| 13x13 | 169 | 6754.218 |
| 17x17 | 289 | 11550.112 |

Using Eq. 9.1, this gives the cell mux space ($M$) values of

$$M_{le} = 5^1 \cdot 4^1 \cdot 2^2 \cdot 2^4 = 2^{10.322} \tag{B.1}$$

for cells S0G-Y, S0F-X and S1F-X, while S1G-Y cells have a mux space of

$$M_{0G} = 4^1 \cdot 4^1 \cdot 2^2 \cdot 2^3 = 2^9. \tag{B.2}$$

Combining these, gives the mux space for a CLB as

$$M_{clb} = M_{le}^3 \cdot M_{0G} = 2^{39.966}. \tag{B.3}$$

Then the size of the evolvable region's state space is calculated by substituting Eq. B.3 into Eq. 9.5 for each of the experiment's CLB matrix sizes. The results of these calculations are given in Table B.1

## B.1.2    Calculating Fitness Feedback

Fitness in the signal routing experiments was based on how much progress was made in routing signal(s), possibly inverted, from the inputs to the outputs, ignoring the relationship between the different inputs and outputs. Thus, the fitness feedback space for the routing problems is based purely on the feedback from the circuit elements, which measure the connectivity of these elements. In other words, the circuit fitness is based solely on circuit structure specifying feedback ($F = F_s$).

The fitness of the circuit is calculated using the connectivity value for each of the logic elements in each of the middle CLB columns, but with only input and output logic elements being used in the input and output columns, respectively. Each logic element's connectivity is measured based on four incrementally applied tests:

1. the LUT input is connected to a neighbour's output;

2. (and) the LUT function passes or inverts its input;

Table B.2: Routing Experiment Fitness Feedback

| Experiment | n | $n'$ | F ($log_2$) | E ($log_2$) |
|---|---|---|---|---|
| 5x5:1-1 | 100 | 62 | 232.193 | 143.960 |
| 9x9:1-1 | 324 | 254 | 752.305 | 589.770 |
| 13x13:1-1 | 676 | 574 | 1569.623 | 1332.787 |
| 17x17:1-1 | 1156 | 1022 | 2684.149 | 2373.011 |
| 8x8:1-1 | 256 | 194 | 594.414 | 450.454 |
| 8x8:4-4 | 256 | 200 | 594.414 | 464.386 |

3. (and) one or more out bus lines is connected to the LUT's output;

4. (and) one of the connected out bus lines has a single line connected to a LUT input in a neighbour.

This gives each logic element 5 (0-4) possible connectivity values ($\mu = 5$) according to how many of the conditions are satisfied. By substituting this into Eq. 9.8 this gives a feedback space of

$$F = 5^n \qquad (B.4)$$

however, noting that the routing experiments are always a square region of CLBs ($r \times r$), and that there are 4 logic elements per CLB, with only one logic element contributing fitness per input ($I$) and output ($O$) at the input/output layers, the number of logic elements contributing to the effective fitness is given by

$$n' = 4r(r - 2) + I + O. \qquad (B.5)$$

and by substituting $n'$ for $n$ in Eq. B.4, the effective feedback space ($E$) can be ascertained with

$$E = 5^{n'}. \qquad (B.6)$$

The values of $n$, $n'$, $F$ and $E$ for each of the routing experiments is supplied in Table B.2.

## B.1.3   Calculating Problem Difficulty

To calculate the difficulty of the routing experiments, using Eq. 9.7, the values of $S$, the total configuration state space, given in Table B.1 can be used, while the values of $A$, the number of configurations that result in the desired configuration, still need to be ascertained.

To determine $A$, it should be noted that the number of solutions is increased by the amount of redundancy created by unused logic elements (that are able to have any configuration without affecting the circuit's function). This means

```
                        Slice

            0 1  0 1  0 1  0 1  0 1
            ------------------------
          | b2.  .  .  .  .  .  . | X
        4 | .  .  .  .  .  .  .  . | Y
          | .  . b4.  . b7 .  . b9. | X
        3 | b3b1 .  .  .  . b8.  .  . | Y
  CLB     | .  . a1b5 .  .  .  .  . yo0 X    Logic
  Row   2 I xi.  .  . b6.  . a5 .  . | Y    Element
          | .  .  . a2 .  . a4.  .  . | X
        1 | .  .  .  . a3.  .  .  .  . | Y
          | .  .  .  .  .  .  .  .  . | X
        0 | .  .  .  .  .  .  .  .  . | Y
            ------------------------

            0    1    2    3    4

                   CLB Column
```

Figure B.1: Shortest Paths on a 5x5 CLB Matrix

that for the signal routing problems, only the shortest path needs to be found to give a first order approximation to the number of solutions (in practical terms they are lost in the rounding of the exponent's decimal places to any reasonable level of precision). Also, it should be noted that as $S1G$-$Y$ logic elements have a smaller configuration space, if two paths with the same length are possible, the path that utilises the most $S1G$-$Y$ logic elements will have the highest redundancy, and hence a higher value of $A$.

### Detailed Approach to Routing Calculations using 5x5 Problem

The two shortest paths for the 5x5 problem are shown in Fig. B.1, with input and output points marked as 'x' and 'y', respectively, and intervening 'a' and 'b' paths numbered.

The steps involved in calculating the value of $A$ for the shortest path (a) are first to determine the number of configurations in the actual path ($\alpha$), that provide the desired solution, and then multiply this by the amount of redundancy supplied by the unused logic elements ($R$), as measured by the total number of configurations these may take.

For the logic elements that contribute to the shortest path, the following configurations are required (note FN=p/i represents a LUT function of pass/invert):

226

- **x**: in=W11, FN=p/i, outbus=OUT1 on, tosingle=OUT1-E3 on;

- **a1**: in=W3, FN=p/i, outbus=OUT2 on, tosingle=OUT2-S5 on;

- **a2**: in=N5, FN=p/i, outbus=OUT3 on, tosingle=OUT3-E11 on;

- **a3**: in=W11, FN=p/i, outbus=OUT1 on, tosingle=OUT1-E3 on;

- **a4**: in=W3, FN=p/i, outbus=OUT2 on, tosingle=OUT2-N8 on;

- **a5**: in=S8, FN=p/i, outbus=OUT4 on, tosingle=OUT4-E14 on;

- **y**: in=W14, FN=p/i, outbus=OUT3 on, tosingle=OUT3-E11 on.

This means that within each contributing logic element there is 1 valid LUT input, 2 valid LUT functions, a contributing out bus with 1 valid setting, a second out redundant bus line (with 2 possible settings), a contributing single line with 1 valid setting (on), and 3 redundant non-contributing single PIPs (each with 2 possible settings). Hence, for each of these logic elements the product of these is taken to give the amount of redundancy per contributing logic element

$$\rho = 1 \cdot 2 \cdot 1 \cdot 2 \cdot 1 \cdot 2^3 = 2^5. \tag{B.7}$$

The number of actively participating components in the path is then given by

$$\alpha = \rho^7 = (2^5)^7 = 2^{35} \tag{B.8}$$

where 7 is the number of participating logic elements in the path.

There are 24 unused logic elements of type *S1G-Y* (which has a smaller configuration space than the other logic elements), and another 69 unused logic elements. This gives the following amount of redundancy

$$R = (5x2^8)^{69} \cdot (2^9)^{24} = 2^{928.213} \tag{B.9}$$

With $\alpha$ and $R$ determined, the number of solutions can be calculated easily, as

$$A = \alpha R = 2^{35} \cdot 2^{928.213} = 2^{963.213}. \tag{B.10}$$

Other paths can be calculated in a similar manner.

Using the results of Eq. B.10 as a first order approximation of $A$ and the value of $S$ given in Table B.1, the difficulty of the problem can be determined using Eq. 9.7 as

$$P = A/S = 2^{963.213}/2^{999.145} = 2^{-35.932}. \tag{B.11}$$

This approach is used to calculate the values of $A$ and $P$ for each of the routing experiments undertaken. The results of these calculations are given in Table B.3

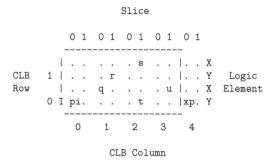

Figure B.2: Shortest Path on a Rx(C+4) CLB Matrix

### Scaling Matrix Size Experiment Calculations

The shortest path solutions for the other scaling routing single input-output experiments (9x9, 13x13, 17x17), utilise the solution for the 5x5, but then prepend multiples of a 4 CLB column matrix, as shown in Fig. B.2, where 'p' is the entry point in the middle CLB row, and 'xp' is either the entry point for the 5x5 solution (shifted to the middle row) for the last 5 CLB columns, or a repeat of the 4 column solution.

The following equation is used to calculate the values of $A$ for each of the 9x9, 13x13, and 17x17 experiments based on the solutions for the 5x5 ($A_{5x5}$ as given by $A$ in Eq. B.10) and 2x4 ($A_{2x4} = 2^{288.439}$), and the CLB mux space $M_{clb}$ given in Eq. B.3:

$$A = A_{5x5} \cdot (A_{2x4})^q \cdot M_{clb}^{(n-(25+8q))} \qquad (B.12)$$

where $q$ is 1, 2 and 3 respectively for the CLB matrix sizes of 9x9, 13x13 and 17x17.

### Scaling IO Experiment Calculations

The shortest path solution used for the 8x8 single input-output experiments, given as CLB row, column coordinates relative to the EHW bounding region, and including slice and logic element is: 3,0,S1F-X; 3,1,S0G-Y; 3,2,S0F-X; 4,2,S1G-Y; 4,3,S1F-X; 4,4,S0G-Y; 4,5,S0F-X; 3,5,S1F-X; 3,6,S0G-Y; 4,6,S1F-X; 4,7,S0G-Y.

The shortest solution to the 4 input 4 output routing problem is shown in Fig. B.3. Note, however, that it is possible to generate solutions that use less logic elements, but involve some inputs not being propagated all the way

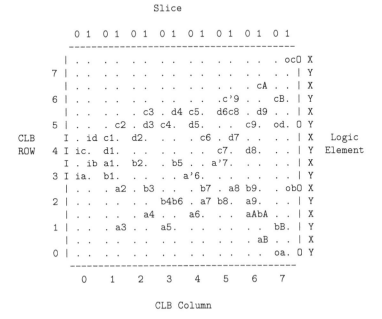

Figure B.3: Shortest Paths on a 8x8 CLB Matrix

to the outputs, while multiple outputs receive the outputs of the same input. Such solutions will still generate a 100% solution so long as the '-forcespread' switch isn't used with the fitness function (as was the case for the experiments presented in this book).

This 'cheat' results in the following path components being used in the optimal solution:

- **Path A**: ia, a1, a'6-a'7, a8-aB, oa.

- **Path B**: ib, b1-bB, ob.

- **Path C**: ic, c1, c'9, cA-cB, oc.

- **Path D**: id, d1-d9, od.

The result of this is that $A$ is increased from $2^{2295}$ to $2^{2351}$; a factor of $2^{56}$.

229

### B.1.4    Summary of Routing Experiment Calculations

Table B.3: Routing Experiment Probabilities in Base 2 Logs

| Experiment | S | F | E | A | P |
|---|---|---|---|---|---|
| 5x5:1-1 | 999.145 | 232.193 | 143.960 | 963.213 | -35.932 |
| 9x9:1-1 | 3237.229 | 752.305 | 589.770 | 3170.010 | -67.219 |
| 13x13:1-1 | 6754.218 | 1569.623 | 1332.787 | 6655.711 | -98.507 |
| 17x17:1-1 | 11550.112 | 2684.149 | 2373.011 | 11420.318 | -129.794 |
| 8x8:1-1 | 2557.810 | 594.414 | 450.454 | 2499.591 | -58.219 |
| 8x8:4-4 | 2557.810 | 594.414 | 464.386 | 2350.897 | -206.913 |

# B.2    Adder Experiment Calculations

## B.2.1    Calculating State Space

Each cell in the unconstrained adder experiments is again mapped to a logic element on the FPGA. This time each logic element supports all four of the LUT inputs, all lines directly connecting from a neighbour (single length lines and special neighbour connecting outputs from some out bus lines), the complete 16-bits of the LUT's functionality, two of the eight out bus lines [1], each of which may either be driven by the unregistered LUT output, or be off, and the six directly connecting (to a neighbouring CLB) single lines driven by each out bus line.

Each logic element thus has the following components, with mux spaces calculated:

- 4 LUT input muxes with 12, 9, 15, 20 states (OFF, input lines), giving $M_{in} = 2^{14.984}$ states;

- 1 LUT with either $M_{lut} = 2^{16}$ (native 16 bit truth table), or $M_{lut} = 2^5$ (estimate for active encoding) states;

- 2 OUT bus lines with 2 states (OFF, asynch LUT output) giving $M_{out} = 2^2$ states;

---

[1]While out bus lines are shared between all cells in a CLB, assigning 2 per cell is used throughout this section for convenience. This has no effect on the calculations as all out bus lines are accounted for. Furthermore, calculations based on routing are determined largely by the settings of the LUT input muxes; the settings of output lines are dependent on the LUT input settings, and as dependent variables are effectively ignored.

Table B.4: Adder Experiment State Spaces in Base 2 Logs

| Experiment | $M_b$ | $S_b$ | $M_a$ | $S_a$ |
|---|---|---|---|---|
| Adder | 44.984 | 719.739 | 33.984 | 543.739 |
| Adder LUT | 16.000 | 256.000 | 5.000 | 38.000 |
| Adder LUT In | 30.984 | 495.739 | 19.984 | 319.739 |
| Adder Mux | 28.984 | 463.739 | 28.984 | 463.739 |

- 12 OutMuxToSingle PIPs (6 per OUT bus line) with 2 states (ON, OFF), giving $M_{pip} = 2^{12}$ states.

Using this information and applying Eq. 9.1, a logic element's mux space can be given as

$$M = M_{in} \cdot M_{lut} \cdot M_{out} \cdot M_{pip}, \qquad (B.13)$$

where in the case of the fixed adder variants, some subset of these will have values of 1, effectively removing them from the equation. Then by combining this with Eq. 9.5, and noting that $n = 16$ for a 2x2 CLB matrix (with the exception of evolving the LUTs only using Active LUTs, in which case only the 8 connected logic elements contribute to the state space), the configuration state space of the evolved region can be found for each of the Adder experiment variants. The results of these calculations are given in Table B.4, with the $b$ and $a$ subscripts being used to denote bit (native or incremental) LUT encoding and active LUT encoding, respectively, and all results given as base-2 logs.

## B.2.2    Calculating Fitness Feedback

For all of the adder experiments, fitness is defined by both circuit function and circuit structure. That is

$$F_{add} = F_s F_f \qquad (B.14)$$

and as the circuit function for an adder is defined by its 3 inputs (x,y,cin) and 2 outputs (sum, cout), substituting these into Eq. 9.9 this gives the functional component of fitness feedback as

$$F_f = 2^{(2^I O)} = 2^{(2^3 \cdot 2)} = 2^{16}. \qquad (B.15)$$

The structural component of fitness is determined by the feedback from the logic elements, and differs between adder experiment variants. For all experiments using a native (Bit) encoding or incremental LUT functions (which are treated equivalently from here on), feedback is provided by a connectivity probe on the LUT's output which tests for any changes in the logic element's output during circuit testing. This results in a connected or not connected status (i.e. 1 bit), so $\mu_{bit} = 2$.

For the adder experiments using Active LUT functions, the connectivity test provides the same amount of feedback as in the signal routing experiments, that being a value equivalent to 0-4, as outlined in Section B.1.2, giving $\mu_{act} = 5$.

By substituting these, along with the number of logic elements (4x4 CLBs = 16 logic elements) into Eq. 9.8, provides the maximum amount of structural feedback $(F_s)$ available for the various adder experiments, these being $2^{16}$ for Bit encoded LUTs and $5^{16} = 2^{37.151}$ for Active encoded LUTs.

So given that $F_f = 2^{16}$, and $F = F_s \cdot F_f$, the maximum amount of feedback available with Bit functions is

$$F_{bit} = 2^{16} \cdot 2^{16} = 2^{32} \tag{B.16}$$

while for Active functions this is

$$F_{act} = 2^{37.151} \cdot 2^{16} = 2^{53.151}. \tag{B.17}$$

**Calculating Effective Fitness Feedback**

To calculate the effective feedback for the adder experiments, first it should be noted that the fitness test for the functional aspect of the adder experiments uses the Hamming distance between desired and actual outputs for each input combination, meaning that all available 16 bits of feedback information was utilised. That is $E_f = 2^{16}$.

On the other hand, as there is only two layers, an input CLB layer (column) and output layer, only those logic elements that are *assigned* as an input or output are included in the connectivity fitness [2], reducing $n = 16$ to $n' = 5$.

The amount of connectivity feedback is further limited to two states for native and incremental LUT encodings, as connectivity is determined solely by probing each LUT's output to determine if any signal changes occur when the input combinations are applied to the circuit's inputs. This is done due to the difficulty of ascertaining how an arbitrary LUT truth table will act on its inputs.

Alternatively, when Active functions are used to encode LUTs, then all non-active lines are ignored by the LUT function, and furthermore only 8 possible basic Boolean functions are able to be applied to these lines (each of which may be inverted or not). This allows a more thorough connectivity analysis to be conducted easily, providing the same degrees of connectivity, per logic element, as the routing experiments. However, when parts of the logic elements are fixed, the effective levels of connectivity that can be differentiated

---

[2]This was an unintended consequence of the re-use of the connectivity analyser used in the signal routing experiments.

is decreased.

When evolving LUTs only (fixing all muxes), Active functions provide no connectivity information, as although on first glance it appears that they would have one bit of information ($\mu' = 2$), that being whether the LUT function passes an active signal, in reality Active LUT functions are guaranteed to pass an active signal for LUTs with active inputs (which all LUTs in the path from inputs to outputs will have, while the other logic elements not connected to the active circuit will pass no signal). Hence, for these experiments $\mu' = 1$.

When only muxes are evolved (with LUT functions fixed), Active function-based logic elements will have four states of connectivity per logic element (i.e. $\mu - 1$, as a valid LUT function is always ensured). Hence in these experiments $\mu' = 4$.

Lastly, when LUTs and their input muxes are evolved (output muxes fixed), feedback is limited to determining whether a LUT input is connected to a neighbour's output, and whether the LUT function passes an active signal. While for Active functions, passing some signal is guaranteed for LUTs with active inputs, those receiving static signals will output static signals. This results in three distinct connectivity states: disconnected, one or more LUT inputs connected to non-active lines, or one or more LUT inputs connected to active lines (hence LUT passes an active signal). So for these experiments, $\mu' = 3$.

The values of $\mu'$ for the different adder experiments is summarised below:

- $\mu' = 1$ for evolving LUT functions with Active LUT encoding;

- $\mu' = 2$ for all experiments with native or incremental LUT encodings;

- $\mu' = 3$ for evolving LUT inputs and function with Active LUT encoding;

- $\mu' = 4$ for evolving muxes with Active LUT encoding;

- $\mu' = 5$ for unconstrained evolution with Active LUT encoding.

Using these values of $\mu'$, with $n' = 5$ and $E_f = 2^{16}$, then the effective fitness for each experiment variant is given as

$$E_{add} = E_s E_f = (\mu')^5 \cdot 2^{16}. \qquad (B.18)$$

The values of $n$, $n'$, $\mu$, $\mu'$, $F$ and $E$ for each of the adder experiments is supplied below in Table B.5.

## B.2.3   Calculating Problem Difficulty

The following sections derive the solutions for the one-bit adder for each experiment, and by locating the solutions that dominate the FPGA configuration

Table B.5: Adder Experiment Fitness Feedback

| Experiment | n | $n'$ | $\mu$ | $\mu'$ | F ($log_2$) | E ($log_2$) |
|---|---|---|---|---|---|---|
| LUTBit | 16 | 5 | 2 | 2 | 32.000 | 21.000 |
| LUTinBit | 16 | 5 | 2 | 2 | 32.000 | 21.000 |
| MuxBit | 16 | 5 | 2 | 2 | 32.000 | 21.000 |
| Bit | 16 | 5 | 2 | 2 | 32.000 | 21.000 |
| LUTAct | 16 | 5 | 5 | 1 | 53.151 | 16.000 |
| LUTinAct | 16 | 5 | 5 | 3 | 53.151 | 23.925 |
| MuxAct | 16 | 5 | 5 | 4 | 53.151 | 26.000 |
| Act | 16 | 5 | 5 | 5 | 53.151 | 27.610 |

space a first-order estimate of the number of solutions can be given. In each case, this is then used to calculate the probability of a random FPGA configuration being a solution for the given experiment.

**Adder Evolving Bit LUTs Only**

Disconnected LUT inputs, and LUT inputs that are undriven, are both held high (at logic level one). This means that for each function of N variables (F4 or F4,F3 or F4,F3,F2 or F4,F3,F2,F1) only the entries marked in Table B.6 with a '*' will ever be used (when routing muxes are fixed, of course).

Noting that a LUT is configured with 16 bits (1 bit per truth table row) the number of configurations for each number of inputs (I) are $2^I$ distinct with $2^{16-I}$ redundant, giving:

- $f()$: there are $2^0$ distinct with $r_{f()} = 2^{16}$ redundant;

- $f(F4)$: there are $2^2$ distinct with $r_{f(F4)} = 2^{14}$ redundant;

- $f(F4, F3)$: there are $2^4$ distinct with $r_{f(F4,F3)} = 2^{12}$ redundant;

- $f(F4, F3, F2)$: there are $2^8$ distinct with $r_{f(F4,F3,F2)} = 2^8$ redundant;

- $f(F4, F3, F2, F1)$: there are $2^{16}$ distinct with $r_{f(F4,F3,F2,F1)} = 2^0$ redundant.

So, remembering that the following functions are required, and noting that pass functions, while comprised of several different alternatives for Active functions, when implemented as a truth table, they are treated as a distinct function, as any other function:

- $3 \times pass(F4)$;

- $2 \times AND(F4, F3)$;

234

Table B.6: Used LUT Boolean Truth Table Entries

```
F4 3 2 1  f(F4) f(F4,F3) f(F4,F3,F2)  f(F4,F3,F2,F1)
------------------------------------------------------
0 0 0 0                                    *
0 0 0 1                        *           *
0 0 1 0                                    *
0 0 1 1         *              *           *
0 1 0 0                                    *
0 1 0 1                        *           *
0 1 1 0                                    *
0 1 1 1  *      *              *           *
1 0 0 0                                    *
1 0 0 1                        *           *
1 0 1 0                                    *
1 0 1 1         *              *           *
1 1 0 0                                    *
1 1 0 1                        *           *
1 1 1 0                                    *
1 1 1 1  *      *              *           *
```

- $2 \times OR(F4, F3)$;

- $1 \times XOR(F4, F3, F2)$.

This means that there are $3 \times f(F4) + 4 \times f(F4, F3) + 1 \times f(F4, F3, F2)$ LUT functions required, so there are:

- 3 $f(F4)$ LUTs with $2^2$ distinct and $2^{14}$ redundant configs;

- 4 $f(F4, F3)$ LUTs with $2^4$ distinct and $2^{12}$ redundant configs;

- 1 $f(F4, F3, F2)$ LUTs with $2^8$ distinct and $2^8$ redundant configs;

of which, for each distinct function there is only 1 valid configuration, and this is increased by the number of redundant LUT configurations $(r)$:

$$
\begin{aligned}
\alpha_{lutbit} &= 1 \cdot r_{f(F4)}^3 \cdot \rho_{f(F4,F3)}^4 \cdot r_{f(F4,F3,F2)}^1 \\
&= 1 \cdot (2^{14})^3 \cdot (2^{12})^4 \cdot (2^8)^1 \\
&= 2^{98}.
\end{aligned}
\tag{B.19}
$$

There are also 8 unused logic elements (LEs)

- 8 $f()$ LUTs with $2^0$ distinct and $2^{16}$ redundant configs.

Which gives

$$\rho_{lutbit} = r_{f()}^8 = (2^{16})^8 = 2^{128}. \tag{B.20}$$

Hence,

$$A_{lutbit} = \alpha_{lutbit} \cdot \rho_{lutbit} = 2^{226}. \tag{B.21}$$

This value of A is for limiting solutions to the configuration:

```
sum  = pass(XOR(x,y,pass(cin)))
     = pass(F4) -> XOR(F4,F3,F2) -> pass(F4)
cout = OR(AND(cin,OR(x,y)),pass(AND(x,y)))
     = AND(F4,F3) ->  pass(F4)  -\ OR(F4,F3)
       OR(F4,F3)  -> AND(F4,F3) -/
```

However, as was the case for Active LUT functions (above), it is necessary to take into account a few variants that can give the same circuit results.

With signal inversions in series *sum* can also be given by (noting that ' ˜ ' is used here to indicate inversion:

```
        ~pass(F4) -> XOR(F4,F3,~F2) -> pass(F4)
    and  pass(F4) -> XNOR(F4,F3,F2) -> ~pass(F4)
```

*cout* can also be given by:

```
        NAND(F4,F3) -> pass(~F4)   -\ OR(F4,F3)
        OR(F4,F3)   -> AND(F4,F3)  -/
    and AND(F4,F3)  -> pass(F4)    -\ OR(F4,F3)
        NOR(F4,F3)  -> AND(F4,~F3) -/
    and AND(F4,F3)  -> ~pass(F4)   -\ OR(F4,~F3)
        OR(F4,F3)   -> AND(F4,F3)  -/
    and NAND(F4,F3) -> pass(F4)    -\ OR(F4,~F3)
        OR(F4,F3)   -> AND(F4,F3)  -/
    and AND(F4,F3)  -> pass(F4)    -\ OR(~F4,F3)
        OR(F4,F3)   -> NAND(F4,F3) -/
    and AND(F4,F3)  -> ~pass(F4)   -\ OR(~F4,~F3)
        OR(F4,F3)   -> NAND(F4,F3) -/
    and NAND(F4,F3) -> pass(F4)    -\ OR(~F4,~F3)
        OR(F4,F3)   -> NAND(F4,F3) -/
```

This gives the following valid configurations for *sum*:

```
        pass(F4) -> XOR(F4,F3,F2) -> pass(F4)   = 1 x 1 x 1 = 1 valid
        ~pass(F4) + XOR(F4,F3,~F2) + pass(F4)   = 1 x 1 x 1 = 1 valid
        pass(F4) -> XNOR(F4,F3,F2) + ~pass(F4)  = 1 x 1 x 1 = 1 valid
```

I.e.

$$\begin{aligned}
\alpha_{sum} &= 3 \cdot r_{f(F4)}^2 \cdot r_{f(F4,F3,F2)}^1 \\
&= 3 \cdot (2^{14})^2 \cdot (2^8)^1 \\
&= 3 \cdot 2^{36} = 2^{37.585}
\end{aligned} \tag{B.22}$$

and the following valid configurations for *cout*:

```
AND(F4,F3) ->  pass(F4)  -\ OR(F4,F3)   = 1 x 1 x 1
 OR(F4,F3) -> AND(F4,F3) -/              x 1 x 1 = 1 valid
 :
variant #8                              = "     "  = 1 valid
```

i.e.

$$\alpha_{cout} = 8 \cdot r^4_{f(F4,F3)} \cdot r^1_{f(F4)}$$
$$= 2^3 \cdot (2^1 2)^4 \cdot (2^{14})^1 = 2^{65} \qquad \text{(B.23)}$$

so the number of possible configurations for the adder is

$$\alpha_{lutbit} = \alpha_{sum} \cdot \alpha_{cout} = 2^{102.585} \qquad \text{(B.24)}$$

then by taking into account the 8 unused LEs, this gives

$$A_{lutbit} = \alpha_{lutbit} \cdot \rho_{lutbit} = 2^{230.585} \qquad \text{(B.25)}$$

and with a search space of $S_{lutbit} = 2^{256}$, the probability of a random configuration of the Bit LUT functions being an adder is:

$$P_{lutbit} = A_{lutbit}/S_{lutbit} = 2^{-25.415}. \qquad \text{(B.26)}$$

**Adder Evolving Active LUTs Only**

There are 8 LUTs used:

```
- 7,11,S0G-Y: XOR(x,y,c)       -> d (s: also dep on fn_c)
- 7,11,S1G-Y: AND(x,y)         -> a (c)
- 7,11,S1F-X: OR(x,y)          -> b (c)
- 7,12,S0G-Y: pass(d)          = m sum: also dep on fn_c,d
- 7,12,S0F-X: pass(a)          -> e (c: also dep on fn_a)
- 8,11,S0G-Y: pass(cin)        -> c (s)
- 8,11,S1F-X: AND(cin,b)       -> f (c: also dep on fn_b)
- 8,12,S1F-X: OR(f,e)          = o cout: also dep on fn_a,b,c,e,f
```

which gives:

```
- sum  = pass(XOR(x,y,pass(cin)))
         dependent on FNs c,d,m
          c -> d -> m
            = pass(F4) -> XOR(F4,F3,F2) -> pass(F4)
- cout = OR(AND(cin,OR(x,y)),pass(AND(x,y)))
         dependent on FNs a,b,c,e,f,o
          a -> e -\ o
```

```
b -> f -/
= AND(F4,F3) ->  pass(F4)  -\ OR(F4,F3)
  OR(F4,F3)  -> AND(F4,F3) -/
```

Noting that *sum* and *cout* are independent of each other's functions.
So have the following LUT functions (note inputs are in order F4,F3,F2,F1):

- 7,11,S0G-Y: $XOR(F4, F3, F2)$

- 7,11,S1G-Y: $AND(F4, F3)$

- 7,11,S1F-X: $OR(F4, F3)$

- 7,12,S0G-Y: $pass(F4)$

- 7,12,S0F-X: $pass(F4)$

- 8,11,S0G-Y: $pass(F4)$

- 8,11,S1F-X: $AND(F4, F3)$

- 8,12,S1F-X: $OR(F4, F3)$

which gives the following LUT functions:

- $3 \times pass(F4)$

- $2 \times AND(F4, F3)$

- $2 \times OR(F4, F3)$

- $1 \times XOR(F4, F3, F2)$

For Active functions, the pass functions can be any of: AND, OR, XOR, MJR0, MJR1; or with inverted inputs: NAND, NOR, XNOR. For the other adder functions, if all inputs are inverted DeMorgan's theorem can be applied, hence the following variants are also possible for *cout*:

```
AND(F4,F3) ->  NOR(~F4,~F3)
OR(F4,F3)  -> NAND(~F4,~F3)
```

Ignoring other combinations of functions that could give the correct outputs, for the moment, then need the following LUT functions (from above):

```
3 x pass(F4)      [ AND(F4), OR(F4), XOR(F4), MJR0(F4), MJR1(F4)
                    NAND(~F4), NOR(~F4), XNOR(~F4) ]
                  -> 8 valid functions
2 x AND(F4,F3)    [or NOR(~F4,~F3)]
                  -> 2 valid functions
2 x OR(F4,F3)     [or NAND(~F4,~F3)]
                  -> 2 valid functions
1 x XOR(F4,F3,F2)
                  -> 1 valid function
```

thus, the number of valid adder configurations is:

$$\alpha_{lutact} = 8^3 \cdot 2^2 \cdot 2^2 \cdot 1 = 2^{13}. \tag{B.27}$$

However, it is also necessary to take into account a few extra solutions that may occur. Knowing that

```
sum  = pass(XOR(x,y,pass(cin)))
     = pass(F4) -> XOR(F4,F3,F2) -> pass(F4)
```

but allowing for signal inversions in series (function output to next stage input mostly), which have no effect on the circuit's function, then the following also produce the correct results:

```
     ~pass(F4) -> XOR(F4,F3,~F2) -> pass(F4)
and  pass(F4) -> XNOR(F4,F3,F2) -> ~pass(F4).
```

Also, knowing that

```
cout = OR(AND(cin,OR(x,y)),pass(AND(x,y)))
     = AND(F4,F3) ->  pass(F4)  -\ OR(F4,F3)
       OR(F4,F3) -> AND(F4,F3) -/
```

and again allowing for signal inversions in series with no effect on the circuit's function, the following variants are possible:

```
      NAND(F4,F3) -> pass(~F4) -\ OR(F4,F3)
      OR(F4,F3) -> AND(F4,F3) -/
and AND(F4,F3) ->  pass(F4)  -\ OR(F4,F3)
      NOR(F4,F3) -> AND(F4,~F3)-/
and AND(F4,F3) -> ~pass(F4)  -\ OR(F4,~F3)
      OR(F4,F3) -> AND(F4,F3) -/
and NAND(F4,F3) -> pass(F4)  -\ OR(F4,~F3)
      OR(F4,F3) -> AND(F4,F3) -/
and AND(F4,F3) ->  pass(F4)  -\ OR(~F4,F3)
      OR(F4,F3) -> NAND(F4,F3) -/
and AND(F4,F3) -> ~pass(F4)  -\ OR(~F4,~F3)
      OR(F4,F3) -> NAND(F4,F3) -/
and NAND(F4,F3) -> pass(F4)  -\ OR(~F4,~F3)
      OR(F4,F3) -> NAND(F4,F3) -/
```

Together, this gives the following valid configurations:

```
sum:
  pass(F4) -> XOR(F4,F3,F2) -> pass(F4)  = 8 x 1 x 8 = 64 configs
  ~pass(F4) + XOR(F4,F3,~F2) + pass(F4)  = 8 x 1 x 8 = 64 configs
  pass(F4) -> XNOR(F4,F3,F2) + ~pass(F4) = 8 x 1 x 8 = 64 configs
```

i.e.

$$a_{sum} = 3 \cdot 64 = 2^{7.585}. \tag{B.28}$$

```
cout:
  AND(F4,F3) ->  pass(F4)  -\ OR(F4,F3) = 2 x 8 x 2
  OR(F4,F3) -> AND(F4,F3) -/            x 2 x 2 = 128 configs
    :
    variant #8                       =  "    "  = 128 configs
```

i.e.

$$a_{cout} = 8 \cdot 2^7 = 2^{10}. \tag{B.29}$$

So the number of possible configurations for the adder is

$$\alpha_{lutact} = a_{sum} \cdot a_{cout} = 2^{17.585}. \tag{B.30}$$

As the unconnected 8 LUTs effectively have a state space of $1$ $(2^0)$, $\rho = 1$, and so

$$A_{lutact} = \alpha_{lutact} \cdot \rho_{lutact} = 2^{17.585} \tag{B.31}$$

and with a search space of $S = 2^{38}$, the probability of a random configuration of the active LUT functions being an adder is:

$$P_{lutact} = A_{lutact}/S_{lutact} = 2^{-20.415} \tag{B.32}$$

### Adder Evolving Bit LUTs and LUT Input Muxes Only

To calculate the number of possible solutions, note that output muxes are fixed (i.e. fixed routing), so the used LUT input muxes are on the one hand restricted to only using the inputs specified by the fixed mux routing (at the input LEs there are several choices due to the input routing CLBs' fan out), but on the other hand, like Active LUT functions, they are not restricted as to which way inputs are ordered.

The following is the valid mux configurations, with mux inputs in F4, F3, F2, F1 order (note, LUT inputs in function description are not ordered here unless marked with a '*' while '%'s mark muxes that are available but would provide the wrong signal to a function):

```
- 7,11,S0G-Y:  XOR(w,w,n*,-);          S0_Y ->  OUT3 -> E11a
   (X): W3,W11;   W6,W8;   %;    W5,OUT_WEST1
   (Y): W14;      W15,W23; %;    W17,W18
    z  #  ;       #;       N12j; #

- 7,11,S1G-Y:  AND(w,w,-,-);          S1_Y -> OUT1h -> E3e
   (X): W5,OUT_WEST1; W2,OUT_WEST0; W6,W8;  W3,W11
   (Y): W17,W18;      W20;          W15,W23; W14

- 7,11,S1F-X:  OR(w,w,-,-);          S1_X -> OUT0 -> N1m
   (X): W5,OUT_WEST1; W2,OUT_WEST0; W6,W8;  W3,W11
```

```
  (Y): W17,W18;        W20;              W15,W23; W14

- 7,12,SOG-Y: pass(w*,-,-,-);              SO_Y ->   OUT1* -> E3* (Sum)
          W11a; #; #; #

- 7,12,SOF-X: pass(w,-,-,-);               SO_X ->   OUT0 -> N1r
          W3e; #; #; OUT_WEST1h

- 8,11,SOG-Y: pass(w,-,-,-);               SO_Y ->   OUT5 -> S12j
    (cin): W3,W11; W6,W8; W2,OUT_WEST0; W5,OUT_WEST1

- 8,11,S1F-X:  AND(w,-,s*,-);              S1_X ->   OUT6 -> E18p
    (cin): W5,OUT_WEST1; W2,OUT_WEST0; %; W3,W11
          #; #; S1m; #

- 8,12,S1F-X:   OR(w*,s*,-,-);             S1_X ->   OUT3* -> E11* (Cout)
          W18p; #; #; #
          #; S1r; #; #
```

LUT input muxes have the following state spaces:

| Logic Element | F4 | F3 | F2 | F1 |
|---|---|---|---|---|
| S0G | 20 | 15 | 9 | 12 |
| S0F | 20 | 15 | 9 | 12 |
| S1G | 12 | 9 | 15 | 20 |
| S1F | 12 | 9 | 15 | 20 |

This is used to determine the valid input mux configurations. However there are overlaps between the possible LUT functions, such as $f(cin, sum)$ and $f(cin, cin, sum)$ etc., and it is the unique LUT configurations that are multiplied by the associated input mux configurations and then these are summed.

Now, due to the difficulty in evaluating the unique elements for each LUT configuration, and the fact that each additional overspecified LUT (such as $f(cin, cin, sum)$) LUT only adds a few extra valid LUT configurations, along with a few input mux configurations, it seems that a first-order approximation would be the most suitable approach for calculating the number of solutions, especially considering that this is only used to calculate a probability to be used for estimating the chances of success.

So what is required is to isolate those LUT functions that are totally independent of each other (i.e. no shared inputs), and each of these is multiplied by the number of mux configurations, and then these are summed. By choosing the LUT function specifications and muxes with the highest redundancy and with no LUT function overlap this will give a first-order estimate.

241

This gives the following configurations, noting that mux inputs are in F4, F3, F2, F1 order, and redundancies are denoted in shorthand by the number of LUT function inputs as $r_{fx}$ rather than $r_{f(..)}$. For example, instead of $r_{f(F4,F3,F2)}$ the shorthand of $r_{f3}$ is used.

```
7,11,SOG-Y:  XOR(w,w,n*,-)
                -,    [2X], n, [2Y]    = 20x2x1x2 = 80
                -,    [2Y], n, [2X]    = 20x2x1x2 = 80
```

$$a = (80 + 80) \times r_{f3} = 160 \times 2^8 = 2^{15.322}$$

```
7,11,S1G-Y:  AND(w,w,-,-)
                [2X], Y,    -,  -      = 2x1x15x20 = 600
                [2Y], [2x], -,  -      = 2x2x15x20 =1200
```

$$a = (600 + 1200) \times r_{f2} = 1800 \times 2^{12} = 2^{22.814}$$

```
7,11,S1F-X:  OR(w,w,-,-)
                [2X], Y,    -,  -      = 2x1x15x20 = 600
                [2Y], [2x], -,  -      = 2x2x15x20 =1200
```

$$a = (600 + 1200) \times r_{f2} = 1800 \times 2^{12} = 2^{22.814}$$

```
7,12,SOG-Y: pass(w*,-,-,-)
                w, -, -, -             = 1x15x9x12 =   1620
```

$$a = 1620 \times r_{f1} = 1620 \times 2^{14} = 2^{24.661}$$

```
7,12,SOF-X: pass(w,-,-,-)
                w, -, -, -             = 1x15x9x12 =   1620
```

$$a = 1620 \times r_{f1} = 1620 \times 2^{14} = 2^{24.661}$$

```
8,11,SOG-Y: pass(w,-,-,-)
                -,  -, [2C],  -        = 20x15x2x12 = 7200
```

$$a = 7200 \times r_{f1} = 7200 \times 2^{14} = 2^{26.814}$$

```
8,11,S1F-X:  AND(w,-,s*,-)
                -, [2C], s, -          = 12x2x1x20 =   480
```

$$a = 480 \times r_{f2} = 480 \times 2^{12} = 2^{20.907}$$

```
8,12,S1F-X:  OR(w*,s*,-,-)
                w, s, -, -             = 1x1x1x1   =    1
```

$a = 1 \times r_{f2} = 1 \times 2^{12} = 2^{12}$

This gives a first order estimate of the total number of solutions:

$$
\begin{aligned}
a_{lutinb} &= 2^{(15.322+22.814+22.814+24.661+24.661+26.814+20.907+12)} \\
&= 2^{169.993}
\end{aligned}
\tag{B.33}
$$

However, it is also necessary to take into account a few variants that can give the same circuit results with signal inversions in series.

*sum* can also be given by:

```
     ~pass(F4) -> XOR(F4,F3,~F2) -> pass(F4)
and  pass(F4) -> XNOR(F4,F3,F2) -> ~pass(F4)
```

while *cout* can also be given by:

```
    NAND(F4,F3) -> pass(~F4) -\ OR(F4,F3)
      OR(F4,F3) -> AND(F4,F3) -/
and AND(F4,F3) ->  pass(F4)  -\ OR(F4,F3)
    NOR(F4,F3) -> AND(F4,~F3)-/

and AND(F4,F3) -> ~pass(F4)  -\ OR(F4,~F3)
     OR(F4,F3) -> AND(F4,F3) -/
and NAND(F4,F3) -> pass(F4)  -\ OR(F4,~F3)
     OR(F4,F3) -> AND(F4,F3) -/
and AND(F4,F3) ->  pass(F4)  -\ OR(~F4,F3)
    OR(F4,F3) -> NAND(F4,F3) -/
and AND(F4,F3) -> ~pass(F4)  -\ OR(~F4,~F3)
    OR(F4,F3) -> NAND(F4,F3) -/
and NAND(F4,F3) -> pass(F4)  -\ OR(~F4,~F3)
    OR(F4,F3) -> NAND(F4,F3) -/
```

For each of these there is a single possible configuration, hence there are 3 variants for sum, and 8 variants for cout. So the number of possible LUT configurations for the adder needs to be multiplied by these:

$$
\alpha_{lutinbit} = a_{lutinb} \cdot 3 \cdot 8 = 2^{174.578}
\tag{B.34}
$$

There are also 8 LEs that don't contribute to the solution, so

$$
\begin{aligned}
\rho_{lutinbit} &= (M_{inmux} \cdot L_{bit})^8 \\
&= (2^{14.984} \cdot 2^{16})^8 = 2^{247.870}
\end{aligned}
\tag{B.35}
$$

this gives the number of solutions as

$$
A_{lutinbit} = \alpha_{lutinbit} \cdot \rho_{lutinbit} = 2^{422.448}
\tag{B.36}
$$

and the probability of a random configuration being a solution as

$$P_{lutinbit} = A_{lutinbit}/S_{lutinbit}$$
$$= 2^{422.448}/2^{495.739} = 2^{-73.292} \tag{B.37}$$

### Adder Evolving Active LUTs and LUT Input Muxes Only

Active LUTs have the same mux values available as Bit LUT functions, however, there won't be all the LUT redundancies, instead there are:

```
3 x pass(F4) [ AND(F4), OR(F4), XOR(F4), MJR0(F4),
               MJR1(F4), NAND(~F4), NOR(~F4), XNOR(~F4) ]
           -> 8 valid functions
2 x AND(F4,F3)    [or NOR(~F4,~F3)]
           -> 2 valid functions
2 x OR(F4,F3)     [or NAND(~F4,~F3)]
           -> 2 valid functions
1 x XOR(F4,F3,F2)
           -> 1 valid function
```

so, the number of valid adder configurations is given by the mux configurations from Bit LUTs, but multiplied by the number of ways of implementing the function (above), giving:

```
7,11,SOG-Y:  XOR(w,w,n*,-)
                 -,    [2X], n, [2Y]   = 20x2x1x2 = 80
                 -,    [2Y], n, [2X]   = 20x2x1x2 = 80
```

$$a = (80 + 80) \times 1 = 160 = 2^{7.322}$$

```
7,11,S1G-Y:  AND(w,w,-,-)
                 [2X], Y,    -,  -     = 2x1x15x20 = 600
                 [2Y], [2x], -,  -     = 2x2x15x20 =1200
```

$$a = (600 + 1200) \times 2 = 3600 = 2^{12.814}$$

```
7,11,S1F-X:  OR(w,w,-,-)
                 [2X], Y,    -,  -     = 2x1x15x20 = 600
                 [2Y], [2x], -,  -     = 2x2x15x20 =1200
```

$$a = (600 + 1200) \times a_{f2} = 1800 \times 2^{12} = 2^{22.814}$$

```
7,12,SOG-Y: pass(w*,-,-,-)
                 w, -, -, -            = 1x15x9x12 =  1620
```

$$a = 1620 \times 8 = 12960 = 2^{13.661}$$

```
7,12,SOF-X: pass(w,-,-,-)
                 w, -, -, -           = 1x15x9x12 =   1620
```

$a = 1620 \times 8 = 12960 = 2^{13.661}$

```
8,11,SOG-Y: pass(w,-,-,-)
                 -,  -, [2C],  -      = 20x15x2x12 = 7200
```

$a = 7200 \times 8 = 57600 = 2^{15.814}$

```
8,11,S1F-X:  AND(w,-,s*,-)
                 -, [2C], s, -        = 12x2x1x20 =   480
```

$a = 480 \times 2 = 960 = 2^{9.907}$

```
8,12,S1F-X:  OR(w*,s*,-,-)
                 w, s, -, -           = 1x1x1x1    =    1
```

$a = 1 \times 2 = 2^1$

This gives a first order estimate of the total number of solutions:

$$
\begin{aligned}
a_{lutina} &= 2^{(7.322+12.814+12.814+13.661+13.661+15.814+9.907+1)} \\
&= 2^{86.993}
\end{aligned}
\tag{B.38}
$$

then taking into account the variants that can give the same circuit results with signal inversions in series (these are the same as for Bit LUTs), there are 3 variants for sum, and 8 variants for cout, so the number of possible LUT configurations for the adder needs to be multiplied by these:

$$
\alpha_{lutinact} = a_{lutina} \cdot 3 \cdot 8 = 2^{91.578}
\tag{B.39}
$$

This needs to be multiplied by the 8 LEs that don't contribute to the solution:

$$
\begin{aligned}
\rho_{lutinact} &= (M_{inmux} \cdot L_{ae})^8 \\
&= (2^{14.984} \cdot 2^5)^8 = 2^{159.870}
\end{aligned}
\tag{B.40}
$$

giving the number of solutions as

$$
A_{lutinact} = \alpha_{lutinact} \cdot \rho_{lutinact} = 2^{251.448}
\tag{B.41}
$$

and the probability of a random configuration being a solution as

$$
\begin{aligned}
P_{lutinact} &= A_{lutinact}/S_{lutinact} \\
&= 2^{251.448}/2^{319.739} = 2^{-68.292}
\end{aligned}
\tag{B.42}
$$

### Adder Evolving Muxes Only (Bit LUT)

With the LUTs' functions fixed, the mux configurations that generate an adder are (LUT inputs are in order 4,3,2,1):

```
- 7,11,S0G-Y:  XOR(w,w,n,-);          S0_Y -> OUT3 -> E11a
                                             OUT7: #
  (X): W3,W11; W6,W8; W2,OUT_WEST0; -
  (Y): W14; W15,W23; W20; -
       N5k,N71; #; N12j; -

- 7,11,S1G-Y:  AND(w,w,-,-);          S1_Y -> OUT1 -> E3e
                                             OUT0: #
  (X): W5,OUT_WEST1; W2,OUT_WEST0; - ; -
  (y): W17,W18; W20; - ; -

- 7,11,S1F-X:  OR(w,w,-,-);           S1_X -> OUT0 -> N1m
                                             OUT1: #
  (X): W5,OUT_WEST1; W2,OUT_WEST0; - ; -
  (y): W17,W18; W20; - ; -

- 7,12,S0G-Y: pass(w,-,-,-);          S0_Y -> OUT1* -> E3* (Sum)
       W11a; - ; - ; -

- 7,12,S0F-X: pass(w,-,-,-);          S0_X -> OUT0 -> N1r
                                             OUT1: #
       W3e; - ; - ; -



- 8,11,S0G-Y: pass(w,-,-,-);          S0_Y -> OUT5 -> S12j
                                             OUT2: S5k, S71
  (cin): W3,W11; - ; - ; -

- 8,11,S1F-X:  AND(w,s,-,-);          S1_X -> OUT6 -> E18p
                                             OUT4: #
  (cin): W5,OUT_WEST1; % ; - ; -
         #; S1m; - ; -

- 8,12,S1F-X:  OR(w,s,-,-);           S1_X -> OUT3* -> E11* (Cout)
       W18p; #; - ; -
       #; S1r; -; -
```

Note that the circuit inputs x, y, cin are routed via

```
x:   7,10,S0G-Y:OUT0-3 - E2,E3,E5,E6,E8,E11
y:   7,10,S1F-X:OUT4-7 - E14,E15,E17,E18,E20,E23
```

```
cin: 8,10,SOG-Y:OUT0-3 - E2,E3,E5,E6,E8,E11
```

and these E lines become the corresponding W lines on the adjacent CLBs (7,11 and 8,11), while OUT0 and OUT1 become OUT_WEST0 and OUT_WEST1.

The '*'s next to the outputs to Sum and Cout are due to the limited CLB outputs that can drive the circuit outputs, these being:

```
Output LE     OUT bus     single
  SOG-Y        OUT1         E3
  S1F-X        OUT3         E11
```

while '#'s indicate that this input, or output doesn't connect to anything (useful), and '%'s are used to indicate that a crossed (wrongly routed) signal can be passed here. Note that crossed signals can occur on:

```
7,12,SOG4: W3;  which would give Sum as AND(x,y)
                instead of XOR(x,y,cin) on W11
7,12,S1F4: W11; which would pass XOR(x,y,cin) to Cout
                instead of AND(x,y)
                resulting in Cout=XOR(x,y,cin) + (x+y).cin
```

Thus, the following muxes can use multiple settings (ignoring the multiple settings allowed on the outputs of 8,11,SOG-Y as the setting required depends on the input used in 7,11,SOG-Y, so it is not an independent variable):

```
- 7,11,SOG-Y: F4=[W3,W11];          F3=[W15,W23];      F2=[N12]
              F4=[W14];             F3=[W6,W8];        F2=[N12]
              F4=[N5,N7];           F3=[W6,W8];        F2=[W20]
              F4=[N5,N7];           F3=[W15,W23];      F2=[W2,OUT_WEST0]
- 7,11,S1G-Y: F4=[W5,OUT_WEST1]; F3=[W20]
              F4=[W17,W18];       F3=[W2,OUT_WEST0]
- 7,11,S1F-X: F4=[W5,OUT_WEST1]; F3=[W20]
              F4=[W17,W18];       F3=[W2,OUT_WEST0]
- 7,12,SOG-Y: F4=[W11]
- 7,12,SOF-x: F4=[W3]
- 8,11,SOG-Y: F4=[W3,W11]
- 8,11,S1F-X: F4=[W5,OUT_WEST1]; F3=[S1]
- 8,12,S1F-x: F4=[W18];          F3=[S1]
```

Ignoring the unused LUT inputs for the moment, this gives the following valid adder circuit combinations (all non-variable settings are ignored, as there is only 1 valid setting).

```
7,11,SOG-Y: 2x2x1 + 1x2x1 + 2x2x1 + 2x2x2 = 18
7,11,S1G-Y: 2x1 + 2x2 = 6
7,11,S1F-X: 2x1 + 2x2 = 6
7,12,SOG-Y: 1
```

```
7,12,S0F-X: 1
8,11,S0G-Y: 2
8,11,S1F-X: 2x2 = 4
8,12,S1F-X: 1x1 = 1
```

This gives the total number of valid mux combos of:

$$a_{mxb} = 18 \cdot 6 \cdot 6 \cdot 1 \cdot 1 \cdot 2 \cdot 4 \cdot 1 = 2^{12.340} \tag{B.43}$$

By taking into account the redundant unused LUT input muxes, this gives the following additional valid combinations:

$$r_{mxb} = 12 \cdot 20 \cdot 15 \cdot 20 \cdot 15 \cdot 12 \cdot 9 \cdot 15 \cdot 12 \cdot 9 \cdot 15 \cdot 12 \cdot 9 \cdot 15 \cdot 20 \cdot 15 \cdot 20 \cdot 15 = 2^{68.486}. \tag{B.44}$$

The redundancy of the "spare", or independent, output lines (single PIPs), driven by the inactive out bus (as each LE provides an output to only one other LUT in the adder) also needs to be taken into account, remembering that only some of these can be utilised without interfering with the redundant LUT inputs, already taken into account previously, and some of these could also cause contention. Hence an estimate of 3 redundant PIPs per LE seems sufficient. Taking this into account for all 8 used LEs, this gives the following extra redundancy

$$r_{oxb} = (2^3)^8 = 2^{24} \tag{B.45}$$

and by including the previous total, this gives the total number of valid adder mux combinations of:

$$\alpha_{muxbit} = a_{mxb} \cdot r_{mxb} \cdot r_{oxb} = 2^{104.825}. \tag{B.46}$$

Then taking into account the 8 unused LEs, and remembering that a LE's mux space, $M_{mux} = 2^{28.984}$, for 8 unused LEs, this gives the redundant mux space as:

$$\rho_{mux} = M_{mux}^8 = 2^{231.872} \tag{B.47}$$

and thus the total adder mux solution space is:

$$A_{muxbit} = \alpha_{muxbit} \cdot \rho_{mux} = 2^{336.872}. \tag{B.48}$$

So, with $S_{muxbit} = 2^{463.739}$ the probability of a random configuration being a solution is

$$P_{muxbit} = A_{muxbit}/S_{muxbit} = 2^{-127.044}. \tag{B.49}$$

### Adder Evolving Muxes Only (Active LUT)

To calculate the number of valid mux settings for fixed ActiveLUT functions, need to look at all the combinations of correct number of inputs on the LUT, coming from appropriate LEs (unlike BitFNs, LUT inputs in function descriptions are not ordered here):

```
7,11,S0G-Y:  XOR(w,w,n,-);          S0_Y ->  OUT3 -> E11a E8b
                                             OUT7: E20c, E23d
  (X): W3,W11;   W6,W8;   W2,OUT_WEST0; W5,OUT_WEST1
  (Y): W14;      W15,W23; W20;          W17,W18
   z   N5k,N7l; #;        N12j;         #


7,11,S1G-Y:  AND(w,w,-,-);           S1_Y ->  OUT1h -> E3e E5f
                                             OUTOi: E2g
  (X): W5,OUT_WEST1; W2,OUT_WEST0; W6,W8; W3,W11
  (y): W17,W18; W20; W15,W23; W14


7,11,S1F-X:  OR(w,w,-,-);            S1_X ->  OUT0 -> N1m N0n
                                             OUT1: N2o
  (X): W5,OUT_WEST1; W2,OUT_WEST0; W6,W8; W3,W11
  (y): W17,W18; W20; W15,W23; W14


7,12,S0G-Y: pass(w,-,-,-);           S0_Y ->  OUT1* -> E3* (Sum)
      W11a; W8b,W23d; W20c; #


7,12,S0F-X: pass(w,-,-,-);           S0_X ->  OUT0 -> N1r N0s
                                             OUT1: N2t
      W3e; #; W2g,OUT_WEST0i; W5f,OUT_WEST1h


8,11,S0G-Y: pass(w,-,-,-);           S0_Y ->  OUT5 -> S12j
                                             OUT2: S5k, S7l
  (cin): W3,W11; W6,W8; W2,OUT_WEST0; W5,OUT_WEST1


8,11,S1F-X:  AND(w,s,-,-);           S1_X ->  OUT6 -> E18p
                                             OUT4: E14q
  (cin): W5,OUT_WEST1; W2,OUT_WEST0; W6,W8; W3,W11
      S2o; S0n; S1m; #


8,12,S1F-X:  OR(w,s,-,-);            S1_X ->  OUT3* -> E11* (Cout)
      W18p; #; #; W14q
      #; S1r; S0s; S2t
```

Thus, the following muxes can use multiple settings (ignoring the multiple settings allowed on the output muxes as the setting required depends on the input used in the receiving CLB, so it is not an independent variable). Noting that this uses the following:

- combination (unordered) of $n$ elements choose $r$: $Cn, r = n!/((n-r)!r!)$

- permutation (ordered) of $n$ elements choose $r$: $Pn, r = N!/(n-r)!$

```
7,11,SOG-Y:
                G4=N combos x P3,2 x %2lines x #linecombos
                + G2=N combos x P3,2 x %2lines x #linecombos
              = 2 x 6 x 5/6 x [2x2]
                + 1 x 6 x 5/6 x [2x2]
              = 3 x 5 x 4
              = 60
7,11,S1G-Y:
                P4,2 x %2lines x #linecombos
              = 12 x 6/8 x [2x2]
              = 12 x 24/8
              = 36
7,11,S1F-X:
                P4,2 x %2lines x #linecombos
              = 12 x 6/8 x [2x2]
              = 12 x 24/8
              = 36
7,12,SOG-Y:
                1 + 2 + 1
              = 4
7,12,SOF-x:
              = 1 + 2 + 2
              = 5
8,11,SOG-Y:
                P4,1 x #linecombos
              = 4 x 2
              = 8
8,11,S1F-X:
                [P3,2 + P3,1] x #linecombos
              = (6 + 3) x [2x1]
              = 18
8,12,S1F-x:
                F4=W combos x P3,1 x #linecombos
                + F1=W combos x P2,1 x #linecombos
              = 1 x 3 x 1
                + 1 x 2 x 1
              = 5
```

This gives the total number of valid mux combinations:

$$a_{mxa} = 60 \cdot 36 \cdot 36 \cdot 4 \cdot 5 \cdot 8 \cdot 18 \cdot 5 = 2^{30.061} \qquad \text{(B.50)}$$

By taking into account the redundant unused LUT inputs (that are unconnected to a driven output), provides the following additional valid combinations. This is an approximation only, as in reality, should take into account the redundancy for each combination of inputs (or choose the one with the

250

highest amount of redundancy). For the cells on the west border, ignore all inputs from the west (input routing CLBs) as these are always driven, but for all other cells assume lines are not driven (as elsewhere are discounting the output line states as not being independent, so here free to choose whatever settings required). This gives

$$r_{mxa} = 8 \cdot 17 \cdot 11 \cdot 17 \cdot 11 \cdot 12 \cdot 9 \cdot 15 \cdot 12 \cdot 9 \cdot 15 \cdot 8 \cdot 6 \cdot 11 \cdot 17 \cdot 11 \cdot 20 \cdot 15 = 2^{64.237} \quad \text{(B.51)}$$

The redundancy of the "spare", or independent, output lines (single PIPs) also needs to be taken into account. As for the Bit LUT encoding, this is estimated to be 3 of the 12 available PIPs per LE. Taking this into account for all 8 used LEs, this gives the following extra redundancy

$$r_{oxa} = (2^3)^8 = 2^{24} \quad \text{(B.52)}$$

Including the previous total, $a_{mxa}$, this gives the total number of valid adder mux combos of:

$$\alpha_{muxact} = a_{mxa} \cdot r_{mxa} \cdot r_{oxa} = 2^{118.298}. \quad \text{(B.53)}$$

For the 8 unused LEs, the redundant mux space is the same as for the case of evolving muxes with Bit LUTs, i.e.

$$\rho_{muxact} = M_{mux}^8 = 2^{231.872}. \quad \text{(B.54)}$$

So the total number of valid adder configurations, including redundant LE configurations is

$$A_{muxact} = \alpha_{muxact} \cdot \rho_{muxact} = 2^{350.170} \quad \text{(B.55)}$$

Hence the probability of a random configuration being a solution is

$$P_{muxact} = A_{muxact}/S_{muxact} = 2^{-113.569}. \quad \text{(B.56)}$$

### Adder Unconstrained Evolution

To get a rough working *estimate* of the number of valid adder configurations ($A_{all}$) when evolving both mux and luts together, start with the number of solutions from multiplexor settings only evolution ($A_{mux}$), where the LUT functions are fixed, and noting that for each function that is needed to generate an adder, there are several possible LUT functions that are able to achieve this, so the initial estimate is multiplied by the number of solutions from LUT only evolution ($A_{lut}$).

In other words the number of possible wiring connection settings for one particular valid set of adder function blocks is multiplied by the number of possible LUT function combinations for one of these particular wirings. Then, to take into account different routing and function combinations that could also work with the increase in logic elements that are utilised, this last estimate is then multiplied by $2^8$, where 8 is the number of extra logic elements that are able to be utilised. This gives

$$A_{all} \approx 2^8 \cdot A_{mux} \cdot A_{lut} \tag{B.57}$$

For native and incremental function encoded LUT approaches, this would be:

$$A_{allbit} \approx 2^8 \cdot 2^{336.695} \cdot 2^{230.585} = 2^{575.280}. \tag{B.58}$$

For active function encoded LUT approaches, it is also necessary to take into account the fact that the other 8 LUTs that weren't used with fixed routing, and had a state space size of 1, could take various configurations, usually without upsetting the circuit's function, so $A_{lut}$ is cubed for good measure (remembering that this is just an estimate). This gives

$$A_{allact} \approx 2^8 \cdot 2^{350.170} \cdot (2^{17.585})^3 = 2^{410.925}. \tag{B.59}$$

These would give *very* rough probability estimates of finding an adder of:

$$P_{allbit} = A_{allbit}/S_{bit} = 2^{575.280}/2^{719.739} = 2^{-144.459} \tag{B.60}$$

and

$$P_{allact} = A_{allact}/S_{act} = 2^{410.925}/2^{543.739} = 2^{-132.814}. \tag{B.61}$$

# B.3    Experiment Calculation Summary

Table B.7: Experiment Calculation Summaries in Base 2 Logs

| Experiment | S | A | P | F | E |
|------------|----------|-----------|----------|----------|----------|
| 5x5:1-1 | 999.145 | 962.213 | -36.932 | 232.193 | 143.960 |
| 9x9:1-1 | 3237.229 | 3170.010 | -67.219 | 752.305 | 589.770 |
| 13x13:1-1 | 6754.218 | 6655.711 | -98.507 | 1569.623 | 1332.787 |
| 17x17:1-1 | 11550.112 | 11420.318 | -129.794 | 2684.149 | 2373.011 |
| 8x8:1-1 | 2557.810 | 2499.591 | -58.219 | 594.414 | 450.454 |
| 8x8:4-4 | 2557.810 | 2349.221 | -208.589 | 594.414 | 464.386 |
| LUTBit | 256.000 | 230.585 | -25.415 | 32.000 | 21.000 |
| LUTinBit | 495.739 | 422.448 | -73.292 | 32.000 | 21.000 |
| MuxBit | 463.739 | 336.695 | -127.044 | 32.000 | 21.000 |
| AllBit | 719.739 | 575.280 | -144.459 | 32.000 | 21.000 |
| LUTAct | 38.000 | 17.585 | -20.415 | 53.151 | 16.000 |
| LUTinAct | 319.739 | 251.448 | -68.292 | 53.151 | 23.925 |
| MuxAct | 463.739 | 350.170 | -113.569 | 53.151 | 26.000 |
| AllAct | 543.739 | 410.925 | -132.814 | 53.151 | 27.610 |

## B.3 Experiment Calculation Summary

Appendix **C**

# Genetic Code Details

This appendix provides details of the genetic encoding of FPGA and simulated resources used in the experiments presented in this book. The manner in which FPGA and simulated resource settings are encoded in the gene coding regions and bind sites of regulatory regions, for the morphogenetic approach to EHW, is presented in Section C.1. Then, in Section C.2 the mapping from intermediate format (*resource, attribute, setting*) to JBits specification, by which manner the subset of available FPGA resources used and their allowed settings is determined, is provided for the morphogenetic approach to both signal routing and adder experiments. Lastly, Section C.3 provides the encodings used by the direct encoding approach for both signal routing and adder experiments.

## C.1 Resource Encoding

The following sections provide information on the top level genetic code (C.1.1) for encoding Virtex FPGA resources and simulated transcription factors (TFs), and the further encoding used by each resource type to specify its setting (Section C.1.2).

The bind code, used for testing if resources bind to bind sites on the chromosome, is based on the same genetic code, and as such, it is only mentioned to the extent that it differs from the genetic code.

### C.1.1 Resource-Attribute Encoding

Table C.1 shows the top level of the genetic code used to encode FPGA and simulated resources by the morphogenetic EHW system. It encodes the first codon (3 sequential bases) of each resource, while the remaining codons required to fully specify a resource and its setting are decoded by resource-specific codes, that are presented in later sections. This table shows the resource type

Table C.1: Top Level Virtex Genetic Code

| 1st | 2nd base | | | | 3rd |
|-----|----------|----------|----------|----------|------|
| base | 0 (U) | 1 (C) | 2 (A) | 3 (G) | base |
| 0 | TF, Local | LUTIncrFN, F | TF, Local | SliceRAM | 0 |
| | TF, Morphogen | LUTIncrFN, G | TF, Morphogen | SliceRAM | 1 |
| | SliceIn, Ctrl | LUTIncrFN, G | STOP | STOP | 2 |
| | SliceIn, Ctrl | LUTIncrFN, F | STOP | SliceIn, LUT | 3 |
| 1 | SliceIn, LUT | SliceToOut, 0 | SliceIn, Ctrl | SliceToOut, 3 | 0 |
| | SliceIn, LUT | SliceToOut, 1 | SliceIn, Ctrl | SliceToOut, 2 | 1 |
| | SliceIn, LUT | SliceToOut, 2 | SliceIn, Ctrl | SliceToOut, 1 | 2 |
| | SliceIn, LUT | SliceToOut, 3 | SliceIn, Ctrl | SliceToOut, 0 | 3 |
| 2 | SliceIn, LUT | LUTIncrFn, G | SliceIn, LUT | LUTBitFn, F | 0 |
| | SliceIn, LUT | LUTIncrFn, F | SliceIn, LUT | LUTBitFn, G | 1 |
| | SliceIn, LUT | LUTIncrFn, F | SliceIn, LUT | LUTBitFn, G | 2 |
| | START / Undef | LUTIncrFn, G | SliceIn, LUT | LUTBitFn, F | 3 |
| 3 | OutToSingleBus, 0 | OutToSingleDir, W | OutToSingleDir, E | OutToSingleBus, 3 | 0 |
| | OutToSingleBus, 1 | OutToSingleDir, S | OutToSingleDir, N | OutToSingleBus, 2 | 1 |
| | OutToSingleBus, 2 | OutToSingleDir, N | OutToSingleDir, S | OutToSingleBus, 1 | 2 |
| | OutToSingleBus, 3 | OutToSingleDir, E | OutToSingleDir, W | OutToSingleBus, 0 | 3 |

and (possibly partial) attribute, shown as *resource, attribute*, that are encoded by each initial resource-specifying codon.

The associated top level bind code, used to determine binding of FPGA resources to the chromosome, is supplied in Table C.2, with *OutToSingle* abbreviated here to *OtoS* for formatting only. The bind code differs from the genetic code, in that it doesn't have gene coding region demarcation codons (start and stop), but has a single *BindSite* control codon, which specifies the start of a bind site, and as with other resource types has its own code, which specifies its own stop codons.

Another difference is the prepended *connect* attribute, used to implement inter-CLB signaling pathways. This attribute means that the resource is located in another CLB, and so it is a connection from another CLB that will be queried for binding. For example a CLB input (*SliceIn*) resource would query the associated CLB out bus to single PIP (*OutToSingleBus* or *OutToSingleDir*). Note that the *connect* attribute-prepended entries are allocated to the connecting resource (not the connected resource).

## Resource Types

The resource-type is used to separate FPGA (and simulated) resources into different categories, these being input muxes (*SliceIn*), CLB slice configuration (*SliceRAM*), LUT function specification (*LUTBitFN, LUTIncrFN, LUTActiveFN*), CLB out bus mux (*SliceToOut*), out bus to single programmable interconnection points (*OutToSingleDir* and *OutToSingleBus*), and simulated transcription factors (*TF*).

There are also a few reserved control codons, these being *START*, which

Table C.2: Top Level Virtex Bind Code

| 1st base | 2nd base | | | | 3rd base |
|---|---|---|---|---|---|
| | 0 (U) | 1 (C) | 2 (A) | 3 (G) | |
| 0 | BindSite | LUTIncrFN, F | BindSite | SliceRAM | 0 |
| | BindSite | LUTIncrFN, G | BindSite | SliceRAM | 1 |
| | SliceIn, connect_Ctrl | LUTIncrFN, G | Undef | Undef | 2 |
| | SliceIn, Ctrl | LUTIncrFN, F | Undef | SliceIn, LUT | 3 |
| 1 | SliceIn, connect_LUT | SliceToOut, 0 | SliceIn, connect_Ctrl | SliceToOut, 3 | 0 |
| | SliceIn, LUT | SliceToOut, 1 | SliceIn, Ctrl | SliceToOut, 2 | 1 |
| | SliceIn, connect_LUT | SliceToOut, 2 | SliceIn, connect_Ctrl | SliceToOut, 1 | 2 |
| | SliceIn, LUT | SliceToOut, 3 | SliceIn, Ctrl | SliceToOut, 0 | 3 |
| 2 | SliceIn, connect_LUT | LUTIncrFn, F | SliceIn, LUT | LUTBitFn, F | 0 |
| | SliceIn, LUT | LUTIncrFn, F | SliceIn, LUT | LUTBitFn, G | 1 |
| | SliceIn, connect_LUT | LUTIncrFn, F | SliceIn, LUT | LUTBitFn, G | 2 |
| | Undef | LUTIncrFn, G | SliceIn, LUT | LUTBitFn, F | 3 |
| 3 | OToSBus, connect_0 | OToSDir, connect_W | OToSDir, connect_E | OToSBus, connect_3 | 0 |
| | OToSBus, 1 | OToSDir, N | OToSDir, N | OToSBus, 2 | 1 |
| | OToSBus, connect_2 | OToSDir, connect_N | OToSDir, connect_S | OToSBus, connect_1 | 2 |
| | OToSBus, 3 | OToSDir, E | OToSDir, W | OToSBus, 0 | 3 |

indicates the start of the gene coding region, and $STOP$ to delimit the end of the gene coding region. If the start codon is encountered within the coding region it is translated as $Undef$ which is ignored.

For experiments that use a single LUT encoding (for example Active LUT functions), a genetic code specification that only contains these LUT encodings is used (for example, with Active functions all LUT entries are specified as $LUTActiveFN$).

If TFs aren't used, these codons may be reallocated (in the experiments presented in Chapter 8, these are reallocated to LUT encoding), or deallocated with $Undef$ (as was done for the experiments in Chapter 6).

Other resources, or subsets of resources, that aren't used in experiments (such as $SliceRAM$) are simply deleted during the conversion from decoded gene to JBits specification process.

## Resource Attributes

The attribute field is used to narrow down from a resource category, to an actual FPGA (or simulated) resource. In the case of LUTs, the attribute specifies which of the two LUTs in the CLB slice is being referred to ($F$ or $G$). TFs, use the attribute field to specify whether they are local to the cell or morphogens, which spread from the original cell to those nearby.

For slice input muxes, CLB out bus muxes, and out bus to single PIPs (programmable interconnection points) the attribute specified with the resource is only partial, the remainder is specified within the resource-specific code.

Table C.3: Resource-Attribute to JBits Correspondence

| Resource | Attribute | JBits Class |
|---|---|---|
| SliceIn | LUT | `F1-F4`, `G1-G4` subclasses of `Mux28To1` |
|  | Ctrl | `BY`,`BX`,`SR`,`CE`,`CLK`,`TS` subclasses of `Mux16To1` |
| LUT*FN | F | `LUT.Slice0_F` |
|  |  | `LUT.Slice1_F` |
|  | G | `LUT.Slice0_G` |
|  |  | `LUT.Slice1_G` |
| SliceRAM |  | `SORAM S1RAM` |
| SliceToOut | 0-7 | `OUT0 - OUT7` subclasses of `Mux12To1` |
| OutToSingleBus | 0-7 | `OutMuxToSingle` |
| OutToSingleDir | N,S,E,W; 0,1 | `OutMuxToSingle` |

**Resource-Attribute to JBits Correspondence**

Table C.3 provides the correspondence between resource-attributes and JBits classes. All classes are found within the `com.xilinx.JBits.Virtex.Bits` package. LUT*FN is used here as shorthand to denote all LUT function encodings (*LUTBitFN*, *LUTIncrFN*, *LUTActiveFN*).

## C.1.2    Resource Settings Encoding

Note that in the following "<" .. ">" is used to denote a codon; <abc> is used to indicate the individual bases (a,b,c) within a codon, and if converted to binary (from base 4) these may be denoted as Aa, Bb, Cc, for example, where the upper case letter indicates the high bit, and lower case indicates low bit extracted from the base 4 digit. Also, to denote that a base within a codon is ignored a period is used, for example <ab.> indicates that the third base is ignored.

A resource's attribute and settings may be denoted as being encoded in a specific codons, such as with <resource/attr-a><attr-b/setting1><setting2>, where "attr-a" and "attr-b" are shorthand for the (parts of the) attribute spread over the resource-encoding and first setting encoding codons, and the last two codons encode the settings. "hi" and "lo" are also used as shorthand for high and low, where appropriate.

Codon numbering is done with the resource/attribute codon numbered as 1, and the following codons numbered sequentially. Alternatively, numbering may be relative to the start of the settings codons only, if explicitly specified.

Note that in the following "Resource", "Attribute(s)" and "Setting(s)" denote parameters supplied to the JBits conversion routines, while "attr" or "attr-a/b/c" are only local to the genetic code. However, while the decode implementation returns a single attribute and a variable number of settings, for passing to the conversion routines, in this section resource-attribute specifies what JBits-defined or simulated resource is being configured, while setting is the setting this is configured to. This division is defined by JBits class and (bits) constant definitions, rather than by the decoding implementation. This approach is used for clarity of hardware configuration, rather than clarity of internal implementation. Nevertheless, the two are equivalent, as the ordering of parameters is the same, only the artificial division between attributes and settings changes.

## LUTActiveFn

A LUTActiveFN encodes the settings for inner classes of LUT in JBits, which are configured with a 16 bit array of (inverted) Boolean truth table values. The LUT's settings are encoded as an expression in the form of *op, inv*, where *op* specifies the Boolean operation (AND, OR, XOR, XNOR, NAND, NOR, MaJoRity0, MaJoRity1) to apply to the active LUT inputs only, and *inv* specifies which inputs (in order 1-4) to invert (0 indicates invert) before applying the Boolean function (to the active inputs).

- Encoding format: $<resource/attr><setting1><setting2>$

- Resource: $LUTActiveFN$

- Attribute: $F/G$

- Settings: *operation, inversion mask*

Settings are encoded as: $<op><inv>$.

- $<op> = <ab.>$: 00 .. 33 (base 4)

- $<inv> = <ab.>$: AaBb (4 bit binary mask l1,l2,l3,l4)

*op* is decoded with Table C.4, giving the Boolean operation to apply to active lines. *op* may be one of AND,OR,XOR,XNOR,NAND,NOR,MJR0,MJR1. MJR0 is majority function with ties broken to 0, and MJR1 is majority function with ties broken to 1. For single active input line expressions, *op* is applied to the same line (for example $AND(l1)$ becomes $AND(l1, l1)$).

*inv* is decoded to a 4 bit mask, l1 - l4, which specifies which input lines are inverted when the Boolean operator is applied. A 0 bit indicates the associated input is to be inverted (if active), while a 1 bit indicates uninverted (if active). The *inv* codon is optional (if codons occur at the end of the coding region for example).

Table C.4: LUT Active and Incremental Functions Operator Encoding

| 1st Base | 2nd base | | | |
|----------|------|------|------|------|
|          | 0    | 1    | 2    | 3    |
| 0        | XNOR | MJR0 | XOR  | MJR1 |
| 1        | AND  | NOR  | NAND | OR   |
| 2        | XOR  | MJR1 | XNOR | MJR0 |
| 3        | OR   | NAND | NOR  | AND  |

**LUTBitFn**

A LUTBitFN encodes the settings for inner classes of LUT in JBits, which are configured with a 16 bit array of (inverted) Boolean truth table values. The LUT's settings are encoded, in Bit functions, as an expression in the form of a 16 bit Boolean truth table.

- Encoding format: $<resource/attr><setting1> .. <setting4>$

- Resource: $LUTBitFN$

- Attribute: $F/G$

- Setting: 16 bit truth table (uninverted)

Settings are encoded in four codons as $<ab.><cd.><ef.><gh.>$ to give $abcdefgh$ (base 4), which is then converted to a 16 bit string $AaBbCcDdEeF$-$fGgHh$.

For example the codon $<213>$ has $21$ extracted, which is then convered to $1001$.

**LUTIncrFn**

A LUTIncrFN encodes the settings for inner classes of LUT in JBits, which are configured with a 16 bit array of (inverted) Boolean truth table values. For Incremental functions, the LUT's settings are encoded as an expression in the form of *set, op, mask*, where *set* specifies how to combine the Boolean operator with the existing LUT configuration (with SET, OR, XOR, AND), *op* specifies the Boolean operation (AND, OR, XOR, XNOR, NAND, NOR, MaJoRity0, MaJoRity1) to apply to the specified LUT inputs, and *mask* is a template that specifies which inputs (in order 1-4) to apply the Boolean function to ('0' or '1', with '-' indicating ignore), and in addition, which of these should be inverted (specified with a '0').

The *set* operator determines the relation between the existing LUT function and the function specified by *op*. SET ignores the existing LUT function, and creates a new LUT function based solely on the function specified by *op* and *mask*, while the other *op* operators specify a bitwise function to combine the LUT's existing truth table with the specified function's truth table.

- Encoding format: *<resource/attr><setting1><setting2><setting3>*

- Resource: *LUTIncrFN*

- Attribute: *F/G*

- Settings: *set, operation, lines mask, inversion mask*

Settings are encoded as: *<set,op><lines><inv>*.

- <set,op> = <abc>: *set*=a, *op*=bc

- <lines> = <ab.>: AaBb (4 bit binary mask l1,l2,l3,l4)

- <inv> = <ab.>: AaBb (4 bit binary mask l1,l2,l3,l4)

*set* is mapped from 0-3 to SET, OR, XOR, and AND operators, in that order.

*op* is mapped from 00 .. 33 (base 4) to the operators defined in Table C.4.

*lines* is decoded to a 4 bit mask, l1 - l4, which specifies which of the selected input lines have the Boolean operator applied. A 1 bit indicates the associated input is used, while a 0 bit indicates the input is ignored.

*inv* is decoded to a 4 bit mask, l1 - l4, which specifies which of the selected input lines is inverted. A 1 bit indicates the associated input is inverted.

The *mask* is created by ANDing the *lines* and *inv* masks, to create a template indicating which lines are utilised in the expression, and which of these are inverted. In the resulting *mask* used by the Incremental function, directly used inputs are denoted with a '1', inverted inputs with a '0', and ignored lines with a '-'.

As an example, *<010><321><032>* is decoded to *set, op=SET, AND; lines=1110* and *inv=0011*. This results in the Incremental LUT expression *SET,AND,110-*.

## OutToSingleBus

*OutToSingleBus* encodes for JBits resource OutMuxToSingle, ordered by out bus line (0-7). There are 6 directly connecting (to the neighbouring CLB's input muxes) single PIPs for each out bus line, each of which may be set to ON or OFF.

- Encoding format: *<resource/attr-a><attr-b/attr-c/setting>*

- Resource: *OutToSingleBus*

Table C.5: Bus – Single Encoding

| 1st Base | 2nd base | | | |
|---|---|---|---|---|
| | 0 | 1 | 2 | 3 |
| 0 | Ignore | | | |
| 1 | lo-1 | lo-2 | hi-1 | hi-2 |
| 2 | lo-3 | lo-4 | hi-3 | hi-4 |
| 3 | lo-5 | lo-6 | hi-5 | hi-6 |

Table C.6: OutToSingleBus Single Line Ordering

| Out Bus | Line Order | | | | | |
|---|---|---|---|---|---|---|
| | 1 | 2 | 3 | 4 | 5 | 6 |
| OUT0 | E2 | N0* | N1* | S1* | S3 | W7 |
| OUT1 | E3 | E5* | N2 | S0* | W4 | W5* |
| OUT2 | E6 | N6 | N8 | S5 | S7 | W9 |
| OUT3 | E8 | E11* | N9 | S10 | W10 | W11* |
| OUT4 | E14 | N12* | N13* | S13* | S15 | W19 |
| OUT5 | E15 | E17* | N14 | S12* | W16 | W17* |
| OUT6 | E18 | N18 | N20 | S17 | S19 | W21 |
| OUT7 | E20 | E23* | N21 | S22 | W22 | W23* |

- Attribute: *0-3 (out bus mod 4), hi/lo 4 bus lines, 1-6 (single line)*

- Setting: *0/1 (ON/OFF)*

The setting is encoded in the second codon, $<abc>$, along with parts of the attribute.

Base $c$ specifies whether the specified PIP is set to ON ($c=1,3$) or OFF ($c=0,2$).

The *attr-a*, supplied with the resource codon, is combined with *attr-b* from the second codon to give an out bus line (0-7), while *attr-c* gives the single line (1-6). Bases $a,b$ from the second codon specify the out bus line and which of its 6 single lines is selected; the decoding of these two bases is shown in Table C.5, with "Ignore" indicating that these encodings aren't used, and hence result in an ignored set of codons.

Table C.6 gives the ordering of the single lines to out bus lines. So that, for example, bus 0 line 5 indicates single S3, which gives the PIP specified in JBits as OUT5_TO_SINGLE_SOUTH3.

Note that the entries in Table C.6 marked with an asterisk are able to cause contention. Specifically, E5/W5 E11/W11 E17/W17 E23/W23 singles can have contention between horizontally adjacent CLBs, and N0/S0 N1/S1

**262**

Table C.7: Out To Single Dir Line Encoding

| 1st Base | 2nd base | | | |
|---|---|---|---|---|
| | 0 | 1 | 2 | 3 |
| 0 | | Ignore | | |
| 1 | 2 | 2 | 3 | 4 |
| 2 | 5 | 6 | 7 | 8 |
| 3 | 9 | 10 | 11 | 12 |

N12/S12 N13/S13 singles can have contention between vertically adjacent CLBs.

## OutToSingleDir

*OutToSingleDir* encodes for JBits resource `OutMuxToSingle`, ordered by destination direction (N,S,E,W). There are 12 directly connecting (to the neighbouring CLB's input muxes) single PIPs for each direction, each of which may be set to ON or OFF.

- Encoding format: *<resource/attr-a><attr-b/setting>*

- Resource: *OutToSingleDir*

- Attributes: *N,S,E,W; 1–12 (single line)*

- Setting: *0/1 (ON/OFF)*

The setting is encoded in the second codon, *<abc>*, along with parts of the attribute.

Base $c$ specifies whether the specified PIP is set to ON (*c=1,3*) or OFF (*c=0,2*).

Bases $a,b$ specify which of the 12 single lines is selected (*attr-b*) in the direction specified by *attr-a*; the decoding of these two bases is shown in Table C.7, with "Ignore" indicating that these encodings aren't used, and hence result in an ignored set of codons.

Table C.8 gives the ordering of the single lines for each direction. So that, for example, North line 5 indicates single N8, which gives the PIP specified in JBits as `OUT2_TO_SINGLE_NORTH8`.

Note that E5/W5 E11/W11 E17/W17 E23/W23 singles can have contention between horizontally adjacent CLBs, and N0/S0 N1/S1 N12/S12 N13/S13 singles can have contention between vertically adjacent CLBs.

Table C.8: Out To Single Dir Single Line Ordering

| Dir | Line Order |
|---|---|
| North | 0-N0 0-N1 1-N2 2-N6 2-N8 3-N9 4-N12 4-N13 5-N14 6-N18 6-N20 7-N21 |
| South | 1-S0 0-S1 0-S3 2-S5 2-S7 3-S10 5-S12 4-S15 6-S17 6-S19 7-S22 |
| East | 0-E2 1-E3 1-E5 2-E6 3-E8 3-E11 4-E14 5-E15 5-E17 6-E18 7-E20 7-E23 |
| West | 1-W4 1-W5 0-W7 2-W9 3-W10 3-W11 5-W16 5-W17 4-W19 6-W21 7-W22 7-W23 |

**SliceIn**

*SliceIn* encodes for connections to slice LUT input muxes F1-F4 and G1-G4 (JBits subclasses of Mux28To1), or slice control muxes BY, BX, SR, CE, CLK, TS (JBits subclasses of Mux16To1).

Each of these muxes may select one of 25 or 26 lines or OFF, however, not all of these lines will be direct CLB outputs, some need to come via a general routing matrix (GRM) switch box, and so for an implementation not manipulating this resource, only direct lines should be used, as was the case for experiments presented in this book.

- Encoding format: *<resource/attr-a><attr-b><setting>*

- Resource: *SliceIn*

- attr-a/b: *LUT/Ctrl; Input Mux (LUT 1-4/ Ctrl 1-6)*

- Attribute: *JBits input mux specification (minus slice prefix)*

- Setting: *Input Line (0-27)*

The attribute defines which slice input mux is being configured, by combining the subset of muxes (LUT or Ctrl) given by *attr-a* with the input mux number (1-4 for LUTs, 1-6 for Ctrl) given by *attr-b*.

*attr-b* is encoded in the first two bases of codon 2 as *<ab.>*. The decoding of these two bases is shown in Table C.9, noting that the *TS* entries were replaced by BX and BY, as the tristate bus wasn't used in the experiments presented in this book.

For example, if *attr-a*=LUT and *<ab.>*=103, this gives the SliceIn attribute of F4 (LUT F line 1); whereas if *attr-a*=Ctrl and *<ab.>*=103, the SliceIn attribute is BX. Noting that in both cases, the attribute will be prepended with the cell's slice number (0 or 1) when converted to JBits format (giving, S0F4 and S0BX for a cell located in slice 0).

The setting, which provides the input line into the selected mux, is encoded in a single codon as *<cde>*, the last base of which is reduced with *e mod 2* to a binary value (i.e. e=0,2 and e=1,3 are treated as the same code),

Table C.9: Slice Input Mux Encoding

| <ab.> | LUT input | Ctrl Input |
|-------|-----------|------------|
| 00 | F1 | CLK |
| 01 | F2 | TS (BX) |
| 02 | F3 | CE |
| 03 | F4 | SR |
| 10 | F4 | BX |
| 11 | F3 | BY |
| 12 | F2 | SR |
| 13 | F1 | CE |
| 20 | G1 | CLK |
| 21 | G2 | TS (BY) |
| 22 | G3 | BX |
| 23 | G4 | BY |
| 30 | G4 | BY |
| 31 | G3 | BX |
| 32 | G2 | CE |
| 33 | G1 | SR |

Table C.10: Slice Input Mux Line Encoding

| <cd(e mod 2)> | Input line number |
|---------------|-------------------|
| 000, 010, (020, 030) | 0, 2 (unused) |
| 001, 011, (021, 031) | 1, 3 (unused) |
| 100, 110, 120, 130 | 4, 6, 8, 10 |
| 101, 111, 121, 131 | 5, 7, 9, 11 |
| 200, 210, 220, 230 | 12, 14, 16, 18 |
| 201, 211, 221, 231 | 13, 15, 17, 19 |
| 300, 310, 320, 330 | 20, 22, 24, 26 |
| 301, 311, 321, 331 | 21, 23, 25, 27 |

giving $4 \cdot 4 \cdot 2 = 32$ distinct codes. However, it is necessary to avoid codons <022><023><032>, so <0><2/3><0-3> are not used, leaving 28 usable codes, which is slightly more than the 25/26+1 input lines available to the muxes, and far more than the available number of direct (from neighbouring CLB) inputs to muxes.

The decoding of the setting codon to an input line number (0–27) is given in Table C.10.

These input line numbers are converted to JBits settings (in the resource, attribute, setting to JBits conversion) using the following scheme. Input line

0=OFF, 1 .. 27 = available lines, unused = no change. The mapping from line number setting to JBits setting is given in Table C.11, with the first entry in the "Inputs Order" column giving line number 0, etc.

As there aren't that many (27) direct lines available, the lines are cycled through, including OFF, however to avoid directional bias, lines are ordered with directions interleaved (for example E1,N1,S1,W1,E2,N2, ..), and the remaining few (less than 4) line numbers are padded with OFF or NOT USED (NOT USED = NOP, i.e. don't do anything, leave CLB as is). Note also that CLB resources may have different lines available according to which slice the mux is located in, for example BY in slice 0 (S0BY) has different lines available to BY in slice 1 (S1BY).

### SliceToOut

*SliceToOut* encodes the settings (X,XB,XQ,Y,YB,YQ,OFF for JBits resources OUT0 – OUT7 (subclasses of Mux12To1) from the appropriate slice. There are 8 out bus muxes, each of which may select any one of the CLB outputs (*S0_X, S0_XB, S0_XQ, S0_Y, S0_YB, S0_YQ, S1_X, S1_XB, S1_XQ, S1_Y, S1_YB, S1_YQ*), the tristate buffer output *TBUF_OUT* or have no input (*OFF*).

- Encoding format: $<resource/attr\text{-}a><attr\text{-}b/setting>$

- Resource: *SliceToOut*

- attr-a/b: *0-3 (out bus mod 4), hi/lo 4 bus lines*

- Attribute: *0-7 (out bus mux number)*

- Setting: *0-7 (slice output)*

The attribute defines which out bus mux is being configured, by combining the out bus mod 4, given by *attr-a* with the high (4-7) or low (0-3) lines specifier given by *attr-b*. For example an *attr-a* of 2 combined with *attr-b* of 'hi' results in a specification of out bus line 6.

*attr-b* is encoded in the first base of codon 2 ($<abc>$), with $a=0,1$ specifying the low 4 bus lines, and $a=2,3$ specifying the high 4 bus lines.

The setting is encoded in bases two and three ($bc$) of codon 2 ($<abc>$), with base $c$ being taken mod 2, to give the slice output (0-7) for configuring the out bus mux. This is shown in Table C.12.

These slice output numbers are converted to JBits settings (in the resource, attribute, setting to JBits conversion) using the following scheme. Slice output 0=OFF, 7 = NOP (reserved for later use for TBUF_OUT), and $1-6$ = available lines (X, XB, XQ, Y, YB, YQ), although where that output isn't available, missing output is padded with a NOP (i.e. ignore). The mapping

Table C.11: Slice Mux Input Line Ordering

| Mux | Inputs order (direct inputs only) |
|---|---|
| S0BY | OFF E7 N5 S0 W8 E19 S8 W14 S9 S14 |
| S1BY | OFF N15 S12 W2 W18 |
| S0BX | OFF E7 N5 S0 W8 E19 S8 W14 S9 S14 |
| S1BX | OFF N15 S12 W2 W18 |
| S0SR | OFF E10 N12 S1 W15 E21 N17 W23 |
| S1SR | OFF E10 N12 S1 W15 E21 N17 W23 |
| S0CE | OFF N19 S6 N22 S21 |
| S1CE | OFF N19 S6 N22 S21 |
| S0F1 | OFF S1_X OUT_WEST1 |
|  | N15 S6 W5 N19 S12 W17 N22 S13 W18 S21 |
| S1F1 | OFF S0_X OUT_EAST6 |
|  | E4 N0 S2 W3 E16 N1 S8 W11 E17 N3 S9 W14 E19 N5 S18 E22 N7 N10 |
| S0G1 | OFF S1_X OUT_WEST1 |
|  | N15 S6 W5 N19 S12 W17 N22 S13 W18 S21 |
| S1G1 | OFF S0_X OUT_EAST6 |
|  | E4 N0 S2 W3 E16 N1 S8 W11 E17 N3 S9 W14 E19 N5 S18 E22 N7 N10 |
| S0F2 | OFF S1_Y OUT_WEST0 |
|  | E11 N12 S1 W2 E23 N17 W20 |
| S1F2 | OFF S0_Y OUT_EAST7 |
|  | E5 N13 S0 W6 E7 S14 W8 E9 S20 W15 E10 W23 E21 |
| S0G2 | OFF S1_Y OUT_WEST0 |
|  | E11 N12 S1 W2 E23 N17 W20 |
| S1G2 | OFF S0_Y OUT_EAST7 |
|  | E5 N13 S0 W6 E7 S14 W8 E9 S20 W15 E10 W23 E21 |
| S0F3 | OFF S0_Y OUT_EAST7 |
|  | E5 N13 S0 W6 E7 S14 W8 E9 S20 W15 E10 W23 E21 |
| S1F3 | OFF S1_Y OUT_WEST0 |
|  | E11 N12 S1 W2 E23 N17 W20 |
| S0G3 | OFF OUT_EAST7 |
|  | E5 N13 S0 W6 E7 S14 W8 E9 S20 W15 E10 W23 E21 |
| S1G3 | OFF OUT_WEST0 |
|  | E11 N12 S1 W2 E23 N17 W20 |
| S0F4 | OFF OUT_EAST6 |
|  | E4 N0 S2 W3 E16 N1 S8 W11 E17 N3 S9 W14 E19 N5 S18 E22 N7 N10 |
| S1F4 | OFF OUT_WEST1 |
|  | N15 S6 W5 N19 S12 W17 N22 S13 W18 S21 |
| S0G4 | OFF S0_X OUT_EAST6 |
|  | E4 N0 S2 W3 E16 N1 S8 W11 E17 N3 S9 W14 E19 N5 S18 E22 N7 N10 |
| S1G4 | OFF S1_X OUT_WEST1 |
|  | N15 S6 W5 N19 S12 W17 N22 S13 W18 S21 |
| S0Clk | OFF GCLK0 GCLK1 GCLK2 GCLK3 N13 |
| S1Clk | OFF GCLK0 GCLK1 GCLK2 GCLK3 N13 |
| TS0 | OFF N0 N3 N10 S18 W3 |
| TS1 | OFF N0 N3 N10 S18 W3 |

Table C.12: Slice Output Mux Line Encoding

| <b(c mod 2)> | Slice output number |
|--------------|---------------------|
| 00 01        | 2, 3                |
| 10 11        | 4, 5                |
| 20 21        | 6, 7                |
| 30 31        | 0, 1                |

Table C.13: Slice Output Mux Line Ordering

| Out Bus Mux | Slice | Outputs order |
|-------------|-------|---------------|
| OUT0        | 0     | OFF S0_X S0_XB S0_XQ S0_Y S0_YB S0_YQ NOP |
|             | 1     | OFF S1_X NOP S1_XQ S1_Y S1_YB S1_YQ NOP |
| OUT1        | 0     | OFF S0_X S0_XB S0_XQ S0_Y S0_YB S0_YQ NOP |
|             | 1     | OFF S1_X NOP S1_XQ S1_Y S1_YB S1_YQ NOP |
| OUT2        | 0     | OFF S0_X S0_XB S0_XQ S0_Y S0_YB S0_YQ NOP |
|             | 1     | OFF S1_X S1_XB S1_XQ S1_Y NOP S1_YQ NOP |
| OUT3        | 0     | OFF S0_X S0_XB S0_XQ S0_Y S0_YB S0_YQ NOP |
|             | 1     | OFF S1_X S1_XB S1_XQ S1_Y NOP S1_YQ NOP |
| OUT4        | 0     | OFF S0_X S0_XB S0_XQ S0_Y NOP S0_YQ NOP |
|             | 1     | OFF S1_X S1_XB S1_XQ S1_Y S1_YB S1_YQ NOP |
| OUT5        | 0     | OFF S0_X S0_XB S0_XQ S0_Y NOP S0_YQ NOP |
|             | 1     | OFF S1_X S1_XB S1_XQ S1_Y S1_YB S1_YQ NOP |
| OUT6        | 0     | OFF S0_X NOP S0_XQ S0_Y S0_YB S0_YQ NOP |
|             | 1     | OFF S1_X S1_XB S1_XQ S1_Y S1_YB S1_YQ NOP |
| OUT7        | 0     | OFF S0_X NOP S0_XQ S0_Y S0_YB S0_YQ NOP |
|             | 1     | OFF S1_X S1_XB S1_XQ S1_Y S1_YB S1_YQ NOP |

from output number setting to JBits setting is given in Table C.13, with the first entry in the "Outputs Order" column giving line number 0, etc.

Note that 022, 023, 032 are stop codons, which is why OFF, NOP and line 6 (which may not be available) are encoded in positions which have some chance of being stop codons. For example low bus lines, line out = OFF can be encoded by 030 (STOP), 032, 130, 132.

## SliceRAM

*SliceRAM* encodes the configuration of the LUTs in the CLB slice, specifying whether they are used as Boolean function generators, shift registers or RAM. It is a higher-level construct that combines the settings for multiple multiplexors.

Table C.14: SliceRAM Attribute-Functionality Encoding

| <ab.> | Attribute | Slice Functionality (Setting) |
|-------|-----------|-------------------------------|
| 00 01 02 03 | DUAL_OFF | SHIFT 16X2 1-LUT SPRAM |
| 10 11 12 13 | DUAL_ON | 2-LUT 2-LUT 2-LUT 2-LUT |
| 20 21 22 23 | DUAL_OFF | SHIFT SHIFT SHIFT SHIFT |
| 30 31 32 33 | DUAL_ON | 2-LUT 32X1 2-LUT DPRAM |

LUT's in a slice can be used as one of:

- 2 4-input lookup tables

- 1 5-input lookup table

- 1 4-1 multiplexor

- 2 16x1-bit synchronous RAM

- 1 16x2-bit synchronous RAM

- 1 16x1-bit dual-port synchronous RAM

- 1 32x1-bit synchronous RAM

This functionality is encoded in 7 fields of the JBits SORAM / S1RAM classes, each of which may be set individually to one of 2 possible settings (ON or OFF). The fields are: DUAL_MODE, LUT_MODE, F_LUT_RAM, G_LUT_RAM, F_LUT_SHIFTER, G_LUT_SHIFTER, RAM_32_X_1. DUAL_MODE has its setting combined with the other field's settings to determine the slice LUT's functionality.

- Encoding format: *<resource><attr/setting>*

- Resource: *SliceToOut*

- attr: -

- Attribute: *DUAL_ON / DUAL_OFF*

- Setting: *LUT/SHIFT/RAM/32X1*

There is a single codon, with only the first two bases used (*<ab.>*) for encoding the attribute and setting. The manner in which these two bases are decoded to provide the 7 different possible slice functionalities is given in Table C.14.

Table C.15: SliceRAM Attribute/Setting to JBits

| Attribute/Setting | JBits Field, Setting |
|---|---|
| DUAL_ON | S0RAM.DUAL_MODE,S0RAM.ON |
| DUAL_OFF | S0RAM.DUAL_MODE,S0RAM.OFF |
| LUT | S0RAM.LUT_MODE,S0RAM.ON |
| RAM | S0RAM.F_LUT_RAM,S0RAM.ON; |
| | S0RAM.G_LUT_RAM,S0RAM.ON |
| 32X1 | S0RAM.RAM_32_X_1,S0RAM.ON |
| SHIFT | S0RAM.F_LUT_SHIFTER,S0RAM.ON; |
| | S0RAM.G_LUT_SHIFTER,S0RAM.ON |

However, note that the returned setting is not the "Slice Functionality", instead, 16x2 is replaced by 32x1, SPRAM and DPRAM are replaced with RAM, and 1-LUT and 2-LUT are replaced by LUT to give the returned setting.

These SliceRAM attributes and settings are converted to JBits settings (in the resource, attribute, setting to JBits conversion) using the scheme in Table C.15, noting that S0RAM is replaced with S1RAM if the cell is located in slice 1, and unspecified fields are all set to OFF.

## TF

TFs encode simulated transcription factors that aren't tied to Virtex (JBits) resources, but bond to portions of DNA to regulate gene transcription. Binding occurs to 'inverted' (0 to 2, 1 to 3) DNA sequences in reverse order. Also, TFs bind only on the 1st and 2nd bases in a codon, the 3rd base is ignored (on both TF and bonding site on chromosome). This is to avoid requiring base sequences on the chromosome (for binding to) that contain start and stop codons - especially start codons in the inhibitory region preceding the start of a gene transcript. This is more of an issue for Virtex-resource based TFs, but to be consistent the same principle is applied here.

- Encoding format: *<resource/attribute><setting1> .. <settingN>*

- Resource: *TF*

- Attribute: *Local/Morphogen*

- Setting: *spread-from-source,bind-string*

TF's are encoded as: TF-type (spread-from-source) binding-sequence. Local TF's are equivalent to Morphogen TF's, except they don't encode the spread from source field, this is set to zero (at source). Note that Cytoplasm type TFs aren't gene-expressible resources, these are only used as cytoplasmic determinants that are set at the start of the developmental process.

Local TFs are encoded as <abc> <def> ...

Morphogen TFs are encoded as: <wx.> <yz.> <abc> <def> ...

the <abc> <def> .. codons are decoded into a binding sequence; while from the <wx.> <yz.> codons used by morphogen type TFs, wxyz encode the spread from source.

*wxyz* is converted (from base-4) to an 8 digit binary string giving the distance to spread from source. This is currently decoded by the morphogenetic EHW system as follows:

- digits 0-3 represent the row spread in Gray code

  - digits 0,1 gives extent of spread to rows in -ve direction

  - digits 2,3 gives extent of spread to rows in +ve direction

- digits 4-7 represent the col spread in Gray code

  - digits 4,5 gives extent of spread to cols in -ve direction

  - digits 6,7 gives extent of spread to cols in +ve direction

The Gray code is 00=0, 01=1, 11=2, 10=3. So, for example *01101100* would be translated as: CLB row spread = from -1 .. +3; CLB col spread = from -2 .. 0.

The binding sequence (*abcdef..*) is encoded directly in the gene, with the sequence terminated on stop codons (022 023 032), which also terminate transcription, or on TF-stop codons (222 223 232) — each of which is a stop codon with the 1st digit 'inverted' from 0 to 2 (in nature U/T and A, C and G are complements, here 0 and 2, 1 and 3) — which only stop production of a TF and not gene transcription. The codons (033 233) surrounded by STOP or stop codons are ignored. Also, to allow TFs to bind on DNA sequences with an odd-count, some codons only code for a single digit, rather than a pair. The decode table is shown in Table C.16.

As an example, the codons <213> <102> <310> <002> <321> <232> decode to: 2 10 31 00 32 (stop). This is reversed, giving: 23 00 13 10 2. And this sequence then binds to the chromosome as in the following.

```
 23.00.13.10.2     TF
 || || || || |
..01x22x31x32x0..  chromosome portion
```

Table C.16: TF Bind Sequence Encoding

| Codon | Bind Sequence |
|-------|---------------|
| 000 010 020 030 | 00 01 02 03 |
| 001 011 021 031 | 00 01 02 03 |
| 002 012 022 032 | 0 0 STOP STOP |
| 003 013 023 033 | 0 0 STOP (unused) |
| 100 110 120 130 | 10 11 12 13 |
| 101 111 121 131 | 10 11 12 13 |
| 102 112 122 132 | 1 1 1 0 |
| 103 113 123 133 | 1 1 1 0 |
| 200 210 220 230 | 20 21 22 23 |
| 201 211 221 231 | 20 21 22 23 |
| 202 212 222 232 | 2 2 stop stop |
| 203 213 223 233 | 2 2 stop (unused) |
| 300 310 320 330 | 30 31 32 33 |
| 301 311 321 331 | 30 31 32 33 |
| 302 312 322 332 | 3 3 3 2 |
| 303 313 323 333 | 3 3 3 2 |

# C.2     Notes on Conversion to JBits Specification

This section provides the mappings from *resource, attribute, setting* to subsets of JBits resources and settings, for the experiments presented in this book. Two different experiment sets were conducted, each of which utilised a different subset of resources. The first of these, detailed in Section C.2.1, was the signal routing experiments, which used a "slim" subset of resources. The second, detailed in Section C.2.2 was the adder experiments, which used a fuller subset, but used only the unregistered logic element outputs.

## C.2.1     Slim Subset for Signal Routing

For the slim routing experiments, logic element cells were used, with each allocated one input mux for its LUT, a subset of available inputs into this mux, the registered logic element output, two out bus muxes, and a subset of the available single PIPs connected to these.

Note that Active LUT functions weren't used for these experiments, and Incremental LUT functions were ignored, and so both are not included here. Also, as the mapping to subsets only affects Virtex resources and their settings, TFs are not affected and so not included. SliceRAM, while included in the genetic code used in the experiments, was ignored (no settings were

272

Table C.17: Slim Subset SliceIn Line Number to Input Line Mapping

| LUT | Line Number mod 5 | | | | |
|-----|------|-----|-----|-----|-----|
| Mux | 0 | 1 | 2 | 3 | 4 |
| S0G4 | OFF | E4 | N7 | S18 | W11 |
| S0F4 | OFF | E19 | N3 | S9 | W3 |
| S1G1 | OFF | E22 | N10 | S8 | - |
| S1F1 | OFF | E16 | N5 | S2 | W14 |

configured), and so it is also not included here.

All numbers, for attributes or settings, referred to in this section are integers.

### SliceIn

The attribute for the *SliceIn* resource specifies the slice input mux. In the slim subset, each LE used only the *S0G4*, *S0F4*, *S1F1* and *S1G1* input muxes. Hence, all LUT input muxes (the resource's attribute) are mapped to this single mux input. For example, *S0G3*, *S0G2* and *S0G1* would all be mapped to *S0G4*. The control input slice muxes *BY*, *BX*, *SR* and *CE* are mapped to the LE's LUT, while *CLK* and *TS* muxes are ignored.

The setting for *SliceIn* is a line number, from 0–27, specifying the input selected for the given input mux. For the slim subset of resources, this is reduced to a line number from 0 to 4 (OFF plus 4 possible inputs) using the modulus operator. For example, given a setting of 17 for an *S0G-X* cell, this would result in a line number of $17 \bmod 5 = 2$.

The mapping from line number (reduced via modulus) to input line for each LUT is given in Table C.17.

### SliceToOut

The attribute for the *SliceToOut* resource specifies the output bus line (*OUT0 – OUT7*) as an integer (0 – 7). Each LE is allocated two out bus muxes, and the mapping from out bus mux number to OUT bus mux for each LE is given in Table C.18.

The setting for SliceToOut resources, specifies the selection of a slice output for a given out bus mux. When using the slim subset of resources, this is limited to OFF or the LE's registered output, only. This is mapped by taking the setting (0–7) and mapping all slice outputs to the single LE output, while the OFF selection remains unchanged, and NOP (setting of 7) is ignored. The setting's mapping is given along with the attribute mapping in Table C.18.

**273**

Table C.18: Slim Subset SliceToOut Out Bus Line and Line Number Mappings

| Logic | Out Bus Number | | Line Number | | |
|-------|-------|-------|-----|-----|-----|
| Element | 0, 2, 4, 6 | 1, 3, 5, 7 | 0 | 1–6 | 7 |
| S0G-Y | OUT0 | OUT1 | OFF | S0_YQ | NOP |
| S0F-X | OUT2 | OUT5 | OFF | S0_XQ | NOP |
| S1G-Y | OUT4 | OUT6 | OFF | S1_YQ | NOP |
| S1F-X | OUT3 | OUT7 | OFF | S1_XQ | NOP |

Table C.19: Slim Subset OutToSingle Out Bus Line and Line Number Mappings

| Logic Element | Out Bus Numbers | Out Bus Mux | Out Single Lines in Line Number Order |
|-------|-------|-------|-------|
| S0G-Y | even | OUT0 | S3 |
|       | odd  | OUT1 | E3, N2, W4 |
| S0F-X | even | OUT2 | N8, S5, S7 |
|       | odd  | OUT5 | W16 |
| S1G-Y | even | OUT4 | E14, W19 |
|       | odd  | OUT6 | N18 |
| S1F-X | even | OUT3 | E11 |
|       | odd  | OUT7 | W22 |

### OutToSingleBus

The (first) attribute for *OutToSingleBus* specifies the out bus number (from 0–7), however, as with the *SliceToOut* resource, these are mapped to the two out bus muxes allocated to that LE.

The second attribute (or from the conversion routine's perspective, the first setting) specifies the output single line (1-6) driven by the specified out bus mux. Due to the restriction on the lines available, the output line number is reduced to fit within the range of the number of lines available, for that out bus mux, using the modulus operator.

The mapping from out bus mux number to OUT bus mux for each LE, and the mapping from reduced output line number to single line is given in Table C.19.

The setting for *OutToSingleBus* is a 0 or a 1, specifying if the Out To Single PIP, given by the attributes above, is turned OFF or ON (respectively).

### OutToSingleDir

The (first) attribute for *OutToSingleDir* specifies the direction (E, N, S, W) for driving a single from the OUT bus, while the second attribute (or from

Table C.20: Slim Subset OutToSingle Out Bus Line and Line Number Mappings

| Logic Element | Dir | OutToSingle Line Order PIPs |
|---|---|---|
| S0G-Y | East | OUT1_TO_SINGLE_EAST3 |
| | North | OUT1_TO_SINGLE_NORTH2 |
| | South | OUT0_TO_SINGLE_SOUTH3 |
| | West | OUT1_TO_SINGLE_WEST4 |
| S0F-X | East | - |
| | North | OUT2_TO_SINGLE_NORTH8 |
| | South | OUT2_TO_SINGLE_SOUTH5, OUT2_TO_SINGLE_SOUTH7 |
| | West | OUT5_TO_SINGLE_WEST16 |
| S1G-Y | East | OUT4_TO_SINGLE_EAST14 |
| | North | OUT6_TO_SINGLE_NORTH18 |
| | South | - |
| | West | OUT4_TO_SINGLE_WEST19 |
| S1F-X | East | OUT3_TO_SINGLE_EAST11 |
| | North | OUT3_TO_SINGLE_NORTH9 |
| | South | OUT3_TO_SINGLE_SOUTH10 |
| | West | OUT7_TO_SINGLE_WEST22 |

the conversion routine's perspective, the first setting) specifies the output line (1-12) driven in the specified direction. Due to the restriction on the lines available, the output line number is reduced to fit within the range of the number of lines available, for that direction, using the modulus operator.

The two attributes are combined together, but with the reduced set of single lines and two dedicated out bus muxes available to the given LE, to specify an OUT To Single PIP. The mapping from direction and line number to OUT To Single PIP is given in Table C.20.

The setting for $OutToSingleDir$ is a 0 or a 1, specifying if the Out To Single PIP, given by the attributes above, is turned OFF or ON (respectively).

## LUTBitFN

The attribute for the $LUTBitFN$ resource specifies the F or G LUT, which is ignored (i.e. it is converted to the appropriate LUT for the given LE) when cells are mapped to LEs, as was done with all experiments presented in this book.

The setting specifies the LUT's function as a binary string of 16 bits, which is the uninverted Boolean truth table for a 4-input LUT. As only one input is used for LUTs in the slim subset of resources, there are only 4 possible functions (always 0, always 1, pass input, invert input), and these can be

Table C.21: Slim Subset LUTBitFN Mapping

| LUT Mux | always 0 00-prefix | pass 01-prefix | invert 10-prefix | always 1 11-prefix |
|---|---|---|---|---|
| S0G4 | 0000000000000000 | 0000000011111111 | 1111111100000000 | 1111111111111111 |
| S0F4 | 0000000000000000 | 0000000011111111 | 1111111100000000 | 1111111111111111 |
| S1G1 | 0000000000000000 | 0101010101010101 | 1010101010101010 | 1111111111111111 |
| S1F1 | 0000000000000000 | 0101010101010101 | 1010101010101010 | 1111111111111111 |

uniquely specified with only the first 2 bits of the LUT's binary string. These mappings are given in Table C.21.

## C.2.2     Unregistered Subset for One Bit Adder

For the Adder experiments, logic element cells were used, comprised of a LUT and its associated input muxes, directly connecting (from neighbouring CLB) input lines, and the unregistered logic element output. The out bus muxes, and the directly connecting (to a neighbouring CLB's LUT) single PIPs connected to these, were shared between all logic elements within a CLB.

Each set of adder experiments used only one LUT function encoding variant, and so all LUT function codons in the genetic code were allocated to the particular variant used for that experiment set (Active, Incremental or Bit). The only JBits conversion done for LUTs was mapping them to the given cell's associated LUT, and so no entries are given for them here.

Note that TFs weren't used for these experiments, and had their codons in the genetic code reallocated to LUT encoding functions. Also, SliceRAM, while included in the genetic code used in the experiments, was ignored (no settings were configured), and so it is also not included here.

**SliceIn**

The attribute for the *SliceIn* resource specifies the slice input mux. In the unregistered subset used by the adder experiments, each LE uses only the LUT input muxes. The control input slice muxes *BY*, *BX*, *SR* and *CE* are mapped to the LE's LUT input muxes (1–4 respectively), while *CLK* and *TS* muxes are ignored.

The setting for *SliceIn* is a line number, from 0–27, specifying the input selected for the given input mux. For the unregistered subset of resources, using directly connecting lines (to neighbouring CLBs) only this is reduced to a line number from 0 to the number of available inputs (including OFF) using the modulus operator. The recurrent CLB lines (such as emphS1_X), which

Table C.22: Unregistered Subset SliceIn Input Line Ordering

| LUT Mux | Line Order |
|---|---|
| S0G1 | OFF, OFF, OUT_WEST1, N15, S6, W5, N19, S12, W17, N22, S13, W18, S21 |
| S0F1 | OFF, OFF, OUT_WEST1, N15, S6, W5, N19, S12, W17, N22, S13, W18, S21 |
| S0G4 | OFF, OFF, OUT_WEST1, N15, S6, W5, N19, S12, W17, N22, S13, W18, S21 |
| S0F4 | OFF, OUT_WEST1, N15, S6, W5, N19, S12, W17, N22, S13, W18, S21 |
| S0G2 | OFF, OFF, OUT_WEST0, E11, N12, S1, W2, E23, N17, W20 |
| S0F2 | OFF, OFF, OUT_WEST0, E11, N12, S1, W2, E23, N17, W20 |
| S1G3 | OFF, OUT_WEST0, E11, N12, S1, W2, E23, N17, W20 |
| S1F3 | OFF, OFF, OUT_WEST0, E11, N12, S1, W2, E23, N17, W20 |
| S0G3 | OFF, OUT_EAST7, E5, N13, S0, W6, E7, S14, W8, E9, S20, W15, E10, W23, E21 |
| S0F3 | OFF, OFF, OUT_EAST7, E5, N13, S0, W6, E7, S14, W8, E9, S20, W15, E10, W23, E21 |
| S1G2 | OFF, OFF, OUT_EAST7, E5, N13, S0, W6, E7, S14, W8, E9, S20, W15, E10, W23, E21 |
| S1F2 | OFF, OFF, OUT_EAST7, E5, N13, S0, W6, E7, S14, W8, E9, S20, W15, E10, W23, E21 |
| S0G4 | OFF, OFF, OUT_EAST6, E4, N0, S2, W3, E16, N1, S8, W11, E17, N3, S9, W14, E19, N5, S18, E22, N7, N10 |
| S0F4 | OFF, OUT_EAST6, E4, N0, S2, W3, E16, N1, S8, W11, E17, N3, S9, W14, E19, N5, S18, E22, N7, N10 |
| S1G1 | OFF, OFF, OUT_EAST6, E4, N0, S2, W3, E16, N1, S8, W11, E17, N3, S9, W14, E19, N5, S18, E22, N7, N10 |
| S1F1 | OFF, OFF, OUT_EAST6, E4, N0, S2, W3, E16, N1, S8, W11, E17, N3, S9, W14, E19, N5, S18, E22, N7, N10 |

cause problems for the VirtexDS simulator, are replaced with an extra OFF entry.

The ordering of input lines (with the first entry being 0) for each LUT input mux is given in Table C.22.

## SliceToOut

The attribute for the *SliceToOut* resource specifies the output bus line (*OUT0 – OUT7*) as an integer (0 – 7), while the setting specifies the selection of a slice output for this out bus mux, given as a line number from 0–7. For the unregistered subset of resources, with cells mapped to logic elements, this results in a setting of line 0 selecting OFF for the specified out bus mux, line 7 (NOP) ignored (reserved for later use for *TBUF_OUT*), and lines 1–6 selecting the logic element's unregistered output (one of *S0_Y, S0_X, S1_Y, S1_X*). [1]

---

[1] Note that there was a minor bug in the conversion used in the adder experiments (discovered after the experiments had been completed), such that a slice granularity mapping

Table C.23: Unregistered Subset OutToSingleBus Line Number Mappings

| Out Bus Number | Out Bus Mux | Out Single Lines in Line Number Order |
|---|---|---|
| 0 | OUT0 | E2, N0, N1, S1, S3, W7 |
| 1 | OUT1 | E3, E5, N2, S0, W4, W5 |
| 2 | OUT2 | E6, N6, N8, S5, S7, W9 |
| 3 | OUT5 | E8, E11, N9, S10, W10, W11 |
| 4 | OUT4 | E14, N12, N13, S13, S15, W19 |
| 5 | OUT6 | E15, E17, N14, S12, W16, W17 |
| 6 | OUT3 | E18, N18, N20, S17, S19, W21 |
| 7 | OUT7 | E20, E23, N21, S22, W22, W23 |

**OutToSingleBus**

The (first) attribute for *OutToSingleBus* specifies the out bus number (from 0–7), while the second attribute (or from the conversion routine's perspective, the first setting) specifies the output single line (1-6) driven by the specified out bus mux.

The mapping from out bus mux number to OUT bus mux for each LE, and the ordering of output lines is given in Table C.23.

The setting for *OutToSingleBus* is a 0 or a 1, specifying if the Out To Single PIP, given by the attributes above, is turned OFF or ON (respectively).

**OutToSingleDir**

The (first) attribute for *OutToSingleDir* specifies the direction (E, N, S, W) for driving a single from the OUT bus, while the second attribute (or from the conversion routine's perspective, the first setting) specifies the output line (1-12) driven in the specified direction.

The two attributes are combined together to specify an OUT To Single PIP. The mapping from direction and line number to OUT To Single PIP is given in Table C.24, noting that PIPs are abbreviated to *outmux_single*. For example, *0_E2* is an abbreviation for *OUT0_TO_SINGLE_EAST2*.

The setting for *OutToSingleDir* is a 0 or a 1, specifying whether the Out To Single PIP, given by the attributes above, is turned OFF or ON (respectively).

---

was used here instead of the logic element granularity that was used by the morphogenesis process. This meant that the that the specified out bus line (0-7) may be configured to take the output from either logic element in the slice. The mapping used was from settings 1-3 to S0_X / S1_X (slice 0 / slice 1 cells) and settings 4-6 to S0_Y / S1_Y, setting 0=OFF, and 7=NOP (i.e. ignored).

Table C.24: Unregistered Subset OutToSingleDir Line Number Mappings

| Dir | No. | OutToSingle PIPs in Line Number Order |
|-----|-----|---------------------------------------|
| E | 1-6 | 0_E2, 1_E3, 1_E5, 2_E6, 3_E8, 3_E11 |
|   | 7-12 | 4_E14, 5_E15, 5_E17, 6_E18, 7_E20, 7_E23 |
| N | 1-6 | 0_N0, 0_N1, 1_N2, 2_N6, 2_N8, 3_N9 |
|   | 7-12 | 4_N12, 4_N13, 5_N14, 6_N18, 6_N20, 7_N21 |
| S | 1-6 | 1_S0, 0_S1, 0_S3, 2_S5, 2_S7, 3_S10 |
|   | 7-12 | 5_S12, 4_S13, 4_S15, 6_S17, 6_S19, 7_S22 |
| W | 1-6 | 1_W4, 1_W5, 0_W7, 2_W9, 3_W10, 3_W11 |
|   | 7-12 | 5_W16, 5_W17, 4_W19, 6_W21, 7_W22, 7_W23 |

# C.3    Notes on Direct Encoding

This section provides the specifications for the encodings used by the direct encoding approach for the experiments presented in this book. For more details on the subsets of resources used, see Sections C.2.1 (for the slim subset used by routing experiments) and C.2.2 (for the unregistered subset used by the adder experiments).

In all cases, the binary chromosome encodes the settings for all logic elements in the evolvable region. Logic elements are ordered starting from the minimum (JBits) CLB row and column. Each row is traversed sequentially in increasing order, within which each column is also traversed in increasing order, and within each CLB the ordering is *S0G-Y*, *S0F-X*, *S1G-Y*, *S1F-X*.

Within each logic element, the configuration is specified by its input mux settings, LUT function, out bus mux settings, and out bus to single mux settings. Two out bus muxes are assigned to each logic element specification, for orthogonality, however, this doesn't prohibit sharing of out bus lines between logic elements.

The number of bits used to encode a given resource's settings is chosen so as to minimise encoding bias, wherever possible. For example, if there are 5 possible settings this can be encoded with 3 bits, however this gives $8 - 5 = 3$ extra settings which must either be ignored, or allocated to 3 of the 5 possible settings. So, in this example, a choice of 4 bits, giving 16 settings with only 1 $(16 - 3 \cdot 5)$ extra, is preferable. When there is an extra setting, this is generally mapped to *OFF* to give a slight bias to off settings.

## C.3.1    Direct Encoding for Signal Routing Experiments

Each logic element is specified as the bits *iiiiffbboooo*. This is decoded as follows:

- *iiii* specifies the input line number for the LUT's input mux;

Table C.25: Direct Encoding Slim Subset Out Bus Assignments and Settings

| Logic | Mux | | Setting | |
|---|---|---|---|---|
| Element | 1 | 2 | 0 | 1 |
| S0G-Y | OUT0 | OUT1 | OFF | S0_YQ |
| S0F-X | OUT2 | OUT5 | OFF | S0_XQ |
| S1G-Y | OUT4 | OUT6 | OFF | S1_YQ |
| S1F-X | OUT3 | OUT7 | OFF | S1_XQ |

- *ff* specifies the LUT function;

- *bb* specifies the out bus selection;

- *oooo* specifies the out bus to single mux settings.

### SliceIn

The input to the single LUT input mux, used in the slim subset of resources, is encoded in 4 bits ($i1$ .. $i4$), giving a line number from 0 to 15. This is then reduced to a number between 0 and 4 for the *S0G4*, *S0F4* and *S1F1* muxes, and from 0 to 3 for *S1G1*.

The mapping from line number (reduced via modulus) to input line for each LUT is given in Table C.17 on page 273.

### LUTBitFN

The LUT's function is applied to only a single input, and hence can be specified with 2 bits, giving 4 possible functions (always 0, always 1, pass input, invert input). The mapping from the encoded two bits (*bb*) to the LUT's full 16 bit binary configuration is given in Table C.21 on page 276.

### SliceToOut

Each logic element is assigned two out bus muxes, as given in Table C.25.

The setting for the out bus muxes is determined by the values of the *bb* bits, with 0 specifying *OFF*, and 1 specifying the registered output of the logic element, also given in Table C.25.

### OutToSingle

Each logic element is allocated 3–4 out to single PIPs, each driven by the out bus lines allocated to that logic element. To encode this, there are 4 bits (*oooo*) assigned for each logic element (in the case of there being only 3 PIPs

Table C.26: Direct Encoding Slim Subset OutToSingle PIPs

| Logic | Bit | | | |
|---|---|---|---|---|
| Element | 1 | 2 | 3 | 4 |
| S0G-Y | 0_S3 | 1_E3 | 1_N2 | 1_W4 |
| S0F-X | 2_N8 | 2_S5 | 2_S7 | 5_W16 |
| S1G-Y | 4_E14 | 4_W19 | 6_N18 | - |
| S1F-X | 3_N9 | 3_S10 | 3_E11 | 7_W22 |

available, the last bit is ignored), with each PIP represented by a single bit ($o$), indicating whether it is set to *OFF* ($o = 0$) or *ON* ($o = 1$). The allocation of bits (1-4) to PIPs for each logic element is given in Table C.26, noting that PIPs are abbreviated to *outmux_single*. For example, *0_S3* is an abbreviation for *OUT0_TO_SINGLE_SOUTH3*.

## C.3.2  Direct Encoding for Adder Experiments

Each logic element is specified as the bits *iiiijjjjkkkkllllffffffffffffffffbbbbccc-cooooooooooooo*. This is decoded as follows:

- *iiii* specifies the input line number (0-11) for the LUT's input mux 1;

- *jjjj* specifies the input line number (0-8) for the LUT's input mux 2;

- *kkkk* specifies the input line number (0-14) for the LUT's input mux 3;

- *lllll* specifies the input line number (0-19) for the LUT's input mux 4;

- *ffffffffffffffff* specifies the LUT function;

- *bbbb* specifies out bus 1 mux selection;

- *cccc* specifies out bus 2 mux selection;

- *oooooooooooo* specifies the out bus to single mux settings.

**SliceIn**

Each of the LUT's input muxes is allocated 4 (or 5) bits, to give a line number from 0 to 15 (or 31), which is then reduced (using the modulus operator) to a line number within the range of the actual number of lines available (including *OFF*).

All the LUT inputs that have the same number of connecting lines are grouped together, so that *iiii* represents one of S0G1/S0F1/S1G4/S1F4 each having 12 available inputs, *jjjj* represents one of S0G2/S0F2/S1G3/S1F3 each having 9 available inputs, *kkkk* represents one of S0G3/S0F3/S1G2/S1F2 each having 15 available inputs, and *lllll* represents one of S0G4/S0F4/S1G1/S1F1 each having 20 available inputs. The ordering of input lines (with the first entry being 0) for each LUT input mux is given in Table C.27.

Table C.27: Direct Encoding Adder SliceIn Input Line Ordering

| Mux | Line Order |
|---|---|
| S0G1/S0F1/S1G4/S1F4 | OFF, N15, S6, W5, N19, S12, W17, N22, S13, W18, S21, OUT_WEST1 |
| S0G2/S0F2/S1G3/S1F3 | OFF, E11, N12, S1, W2, E23, N17, W20, OUT_WEST0 |
| S0G3/S0F3/S1G2/S1F2 | OFF, E5, N13, S0, W6, E7, S14, W8, E9, S20, W15, E10, W23, E21, OUT_EAST7 |
| S0G4/S0F4/S1G1/S1F1 | OFF, E4, N0, S2, W3, E16, N1, S8, W11, E17, N3, S9, W14, E19, N5, S18, E22, N7, N10, OUT_EAST6 |

**LUTBitFN**

The LUT function is completely specified with the 16 bit (uninverted) binary truth table given by *ffffffffffffffff*.

**SliceToOut**

Each logic element is assigned two out bus muxes (the same as was done for the signal routing experiment encoding); see Table C.25 on page 280 for details.

Here, however, the out bus muxes are able to select any of the CLB's 4 unregistered outputs and *OFF*. This means that there are 5 possible settings for each out bus mux. The setting for the out bus muxes are determined by the values of the *bbbb* bits for the first bus and *cccc* for the second. This gives 16 possible settings, with the extra setting $(16 - (3 \cdot 5) = 1)$ assigned to *OFF*. The 4 bits are decoded from binary (*bbbb* or *cccc*) to decimal (0-15), and then reduced to a value from 0-4 with the modulus operator. The ordering, starting from 0, for the mux settings is: *OFF, S0_Y, S0_X, S1_Y, S1_X*.

**OutToSingle**

Each logic element is allocated 12 out to single PIPs, each driven by the out bus lines allocated to that logic element. To encode this, there are 12 bits (*oooooooooooo*) assigned for each logic element, with each PIP represented by a single bit (*o*), indicating whether it is set to *OFF* ($o = 0$) or *ON* ($o = 1$). The ordering of the allocation of bits to PIPs, for each logic element, is given in

Table C.28, noting that PIPs are abbreviated to *outmux_single*. For example, *0_E2* is an abbreviation for *OUT0_TO_SINGLE_EAST2*.

Table C.28: Direct Encoding Unregistered Subset OutToSingle PIPs

| LE | Bits 1-12 |
|---|---|
| S0G-Y | 0_E2, 0_N0, 0_N1, 0_S1, 0_S3, 0_W7, 1_E3, 1_E5, 1_N2, 1_S0, 1_W4, 1_W5 |
| S0F-X | 2_E6, 2_N6, 2_N8, 2_S5, 2_S7, 2_W9, 5_E15, 5_E17, 5_N14, 5_S12, 5_W16, 5_W17 |
| S1G-Y | 4_E14, 4_N12, 4_N13, 4_S13, 4_S15, 4_W19, 6_E18, 6_N18, 6_N20, 6_S17, 6_S19, 6_W21 |
| S1F-X | 3_E8, 3_E11, 3_N9, 3_S10, 3_W10, 3_W11, 7_E20, 7_E23, 7_N21, 7_S22, 7_W22, 7_W23 |

# Bibliography

Aggarwal, V. (2003). Evolving sinusoidal oscillators using genetic algorithms. In *The 2003 NASA/DoD Conference on Evolvable Hardware*, pp. 67–76. IEEE Computer Society.

Albert, D. (1997). Evolutionary hardware overview. http://citeseer.nj.nec.com/albert97evolutionary.html.

Anadigm Inc. (2002). AN10E40 field programmable analog array. http://www.anadigm.com.

Anderson, P. (1998, May). Evolvable hardware: Artificial evolution of hardware circuits in simulation and reality. Master's thesis, Department of Computer Science, University of Aarhus, Ny Munkegade, 8000 Aarhus C.

Banzhaf, W. (1994). Geonotype-phenotype-mapping and neutral variation – a case study in genetic programming. In Y. Davidor, H.-P. Schwefel, and R. Manner (Eds.), *Parallel Problem Solving from Nature (PPSN III)*, pp. 322–332. Springer.

Bentley, P. (2003, March 17–20). Evolving fractal proteins. In A. M. Tyrrell, P. C. Haddow, and J. Torresen (Eds.), *Proceedings of the 5th International Conference on Evolvable Systems: From Biology to Hardware ICES 2003*, Volume 2606 of *Lecture Notes in Computer Science*, Trondheim, Norway, pp. 81–92. Springer.

Bentley, P. (2004). Adaptive fractal gene regulatory networks for robot control. In J. Miller (Ed.), *Proceedings of the Workshop on Regeneration and Learning in Developmental Systems, Genetic and Evolutionary Computation Conference (GECCO2004)*.

Bentley, P., T. Gordon, J. Kim, and S. Kumar (2001). New trends in evolutionary computation. In *Proceedings of the 2001 Congress on Evolutionary Computation*, Volume 1, pp. 162–169.

Bentley, P. and S. Kumar (1999). Three ways to grow designs: A comparison of evolved embryogenies for a design problem. In *Proceedings of the*

*Genetic and Evolutionary Conference (GECCO '99)*, pp. 35–43.

Berenson, D., N. Estévez, and H. Lipson (2005, July 1–3). Hardware evolution of analog circuits for in-situ robotic fault-recovery. In *Proceedings of the 2005 NASA/DoD Conference on Evolvable Hardware*, Washington D.C., USA, pp. 12–19. IEEE Computer Society.

Bradley, D., C. Ortega-Sanchez, and A. Tyrell (2000). Embryonics + Immunotronics: A bio-inspired approach to fault tolerance. In J. Lohn et al. (Eds.), *Proceedings of the Second NASA/DoD Workshop on Evolvable Hardware*, pp. 215–233. IEEE Computer Society.

Bradley, D. and A. Tyrell (2000a, October). Hardware fault tolerance: An immunological solution. In *Proceedings of the IEEE Conference on Systems, Man and Cybernetics*, Volume 1, , Nashville, USA, pp. 107–112.

Bradley, D. and A. Tyrell (2000b, April). Immunotronics: Hardware fault tolerance inspired by the immune system. In J. Miller, A. Thompson, P. Thomson, and T. Fogarty (Eds.), *Proceedings of the Third International Conference on Evolvable Systems (ICES2000)*, Lecture Notes in Computer Science, 1801, pp. 11–20. Springer-Verlag.

Bradley, D. and A. Tyrell (2001, July). The architecture for a hardware immune system. In *Proceedings of the Third NASA/DoD Workshop on Evolvable Hardware*, Long Beach, California, USA, pp. 193–200.

Burke, D. S., K. A. De Jong, J. J. Grefenstette, C. L. Ramsey, and A. S. Wu (1998). Putting more genetics into genetic algorithms. *Evolutionary Computation 6*(4), 387–410.

Carmichael, C. (1999, March 21). Virtex configuration and readback (version 1.0). Technical Report Application Note: XAPP 138, Xilinx.

Channon, A. and R. Damper (1998, June 26-29). Evolving novel behaviors via natural selection. In C. Adami, R. Belew, H. Kitano, and C. Taylor (Eds.), *Proceedings of Artificial Life VI*, Los Angeles, pp. 384–388. MIT Press.

Clark, G. R. (1999). A novel function-level EHW architecture within modern FPGAs. In *Proceedings of the 1999 Congress on Evolutionary Computation (CEC99)*, Volume 2, pp. 830–833.

Crick, F. (1966). Codon-anticodon pairing; the wobble hypothesis. *Journal of Molecular Biology 19*, 548–555.

Darrin, A. G., R. Conde, B. Chern, P. Luers, S. Jurczyk, and C. Mills (2001, July). Adaptive instrument module: Space instrument controller "brain" through programmable logic devices. In *Proceedings of the Third NASA/DoD Workshop on Evolvable Hardware*, Long Beach, California, USA, pp. 256–260.

de Garis, H. (1999). Artificial embryology and cellular differentiation. In P. Bentley (Ed.), *Evolutionary Design by Computers*, pp. 281–295. Morgan Kaufmann: San Francisco, CA.

de Garis, H. and M. Korkin (2002, February). The cam-brain machine (cbm) an FPGA based hardware tool which evolves a 1000 neuron net circuit module in seconds and updates a 75 million neuron artificial brain for real time robot control. *Neurocomputing 42*(1–4).

Eggenberger, P. (1996). Cell interactions as a control tool of developmental processes for evolutionary robotics. In *Proceedings of SAB '96*, pp. 440–448.

Eggenberger, P. (1997a, October 8-10). Creation of neural networks based on developmental and evolutionary principles. In *Proceedings of the International Conference on Artificial Neural Networks (ICANN'97)*, Lausanne, Switzerland.

Eggenberger, P. (1997b). Evolving morphologies of simulated 3d organisms based on differential gene expression. In P. Husbands and I. Harvey (Eds.), *FOURTH EUROPEAN CONFERENCE ON ARTIFICIAL LIFE (ECAL97)*. MIT Press.

Flockton, S. J. and K. Sheehan (1998). Intrinsic circuit evolution using programmable analogue arrays. In *Second International Conference on Evolvable Systems: from biology to hardware (ICES98), Lecture Notes in Computer Science*, Volume 1478, pp. 144–153. Springer.

Flockton, S. J. and K. Sheehan (2000, July). Behaviour of a building block for intrinsic evolution of analogue signal shaping and filtering circuits. In J. Lohn et al. (Eds.), *The Second NASA/DoD Workshop on Evolvable Hardware*. IEEE Computer Society.

Fogarty, T., J. Miller, and P. Thomson (1997). Evolving digital logic circuits on xilinx 6000 family FPGAs. In *The Second Online Conference on Soft Computing*.

Friedman, G. J. (1956). Selective feedback computers for engineering synthesis and nervous system analogy. Master's thesis, UCLA.

Garvie, M. and A. Thompson (2003). Evolution of combinatorial and sequential on-line self-diagnosing hardware. In *The 2003 NASA/DoD Conference on Evolvable Hardware*, pp. 167–173. IEEE Computer Society.

Garvie, M. and A. Thompson (2004). Scrubbing away transients and jiggling around the permanent: Long survival of FPGA systems through evolutionary self-repair. In C. Metra, R. Leveugle, M. Nicolaidis, and J. Teixeira (Eds.), *Proc. 10th IEEE Intl. On-Line Testing Symposium*, pp. 155–160. IEEE Computer Society.

Gilbert, S. (1997). *Developmental Biology* (Fifth ed.). Sunderland, MA: Sinauer Associates.

Gordon, T. (2003, December 8–12). Exploring models of development for evolutionary circuit design. In *Congress on Evolutionary Computation CEC2003*, Volume 3, Canberra, Australia, pp. 2050–2057. IEEE Press.

Gordon, T. G. and P. J. Bentley (2002a). On evolvable hardware. In S. Ovaska and L. Sztandera (Eds.), *Soft Computing in Industrial Electronics*, pp. 279–323. Physica-Verlag, Heidelberg, Germany.

Gordon, T. G. and P. J. Bentley (2002b, July 15-18). Towards development in evolvable hardware. In *Proceedings of the 2002 NASA/DoD Conference on Evolvable Hardware*, Washington D.C., USA.

Gordon, T. G. and P. J. Bentley (2005, July 1–3). Development brings scalability to hardware evolution. In *Proceedings of the 2005 NASA/DoD Conference on Evolvable Hardware*, Washington D.C., USA, pp. 272–279. IEEE Computer Society.

Green, D. G. (1993). L-systems tutorial. http://life.csu.edu.au/complex/tutorials/ tutorial2.html.

Gruau, F. (1994). *Neural network synthesis using cellular encoding and the genetic algorithm*. Ph. D. thesis, Ecole Normale Superieure de Lyon.

Gruau, F. and K. Quatramaran (1996). Cellular encoding for interactive evolutionary robotics. Technical Report 425, University of Sussex.

Guccione, S., D. Levi, and P. Sundararajan (1999, September). Jbits: Java based interface for reconfigurable computing. In *Second Annual Military and Aerospace Applications of Programmable Devices and Technologies Conference (MAPLD)*, Laurel, MD.

Gwaltney, D. A. and M. I. Ferguson (2003). Intrinsic hardware evolution for the design and reconfiguration of analog speed controllers for a dc motor. In *The 2003 NASA/DoD Conference on Evolvable Hardware*, pp. 81–90. IEEE Computer Society.

Gwaltney, D. A. and M. I. Ferguson (2005, July 1–3). Enabling the online intrinsic evolution of analog controllers. In *Proceedings of the 2005 NASA/DoD Conference on Evolvable Hardware*, Washington D.C., USA, pp. 3–11. IEEE Computer Society.

Haddow, P. C. and G. Tufte (1999). Evolving a robot controller in hardware. In *Proceedings of NIK'99*, pp. 141–150.

Haddow, P. C. and G. Tufte (2000). An evolvable hardware FPGA for adaptive hardware. In *Proceedings of the 2000 Congress on Evolutionary Computation (CEC '00)*, Piscataway, NJ, pp. 553–560.

Haddow, P. C. and G. Tufte (2001, July). Bridging the genotype-phenotype mapping for digital FPGAs. In D. Keymeulen, A. Stoica, and J. Lohn (Eds.), *Proceedings of the Third NASA/DoD Workshop on Evolvable Hardware*, Long Beach, California, USA, pp. 109–115.

Haddow, P. C., G. Tufte, and P. van Remortel (2001, October 3-5). Shrinking the genotype: L-systems for ehw? In *Proceedings of the Fourth International Conference on Evolvable Systems: From Biology to Hardware (ICES01). Lecture Notes in Computer Science*, Volume 2210, Tokyo, Japan. Springer.

Hadžić, I., S. Udani, and J. M. Smith (1999). FPGA viruses. In *Proceedings of FPL*, pp. 291–300.

Harvey, I. and A. Thompson (1996). Through the labyrinth evolution finds a way: A silicon ridge. In T. Higuchi (Ed.), *Proceedings of the First International Conference on Evolvable Systems: From Biology to Hardware (ICES96)*, pp. 406–422. Springer-Verlag.

Higuchi, T., H. Iba, and B. Manderic (1994). Evolvable hardware. In H. Kitano (Ed.), *Massively Parallel Artificial Intelligence*. MIT Press.

Higuchi, T., M. Iwata, I. Kajitani, H. Yamada, B. Manderick, Y. Hirao, M. Murakawa, S. Yoshizawa, and T. Furuya (1996). Evolvable hardware with genetic learning. In *IEEE International Symposium on Circuits and Systems (ISCAS '96)*, Volume 4, pp. 29–32.

Higuchi, T., M. Iwata, D. Keymeulen, H. Sakanashi, M. Murakawa, I. Kajitani, E. Takahashi, K. Toda, M. Salami, N. Kajihara, and N. Otsu (1999, September). Real-world applications of analog and digital evolvable hardware. *IEEE Transactions on Evolutionary Computation 3*(3), 220–235.

Higuchi, T., M. Iwata, E. Takahashi, Y. Kasai, H. Sakanashi, M. Murakawa, and I. Kajitani (2000). Development of evolvable hardware at electrotechnical laboratory. In *Proceedings of the 26th Annual Conference of IEEE Industrial Electronics Society (IECON2000)*, Volume 4, pp. 2981–2985.

Higuchi, T., M. Murakawa, M. Iwata, I. Kajitani, W. Liu, and M. Salami (1997). Evolvable hardware at function level. In *IEEE International Conference on Evolutionary Computation*, pp. 187–192.

Hollingworth, G., S. Smith, and A. Tyrell (2000). Safe intrinsic evolution of virtex devices. In J. Lohn et al. (Eds.), *The Second NASA/DoD Workshop on Evolvable Hardware*. IEEE Computer Society.

Hollingworth, G., A. Tyrell, and S. Smith (1999, May 26-27). Simulation of evolable hardware to solve low level image processing tasks. In

R. Poli, H.-M. Voigt, S. Cagnoni, D. Corne, G. Smith, and T. C. Fogarty (Eds.), *EvoWorkshops: Proceedings of Evolutionary Image Analysis, Signal Processing and Telecommunications, First European Workshops, EvoIASP'99 and EuroEcTel'99*, Volume 1596 of *Lecture Notes in Computer Science*, Göteborg, Sweden, pp. 46–58. Springer.

Hong, J.-H. and S.-B. Cho (2003, December 8–12). Meh: Modular evolvable hardware for designing complex circuits. In *Congress on Evolutionary Computation CEC2003*, Volume 1, Canberra, Australia, pp. 92–99. IEEE Press.

Hornby, G. S., H. Lipson, and J. B. Pollack (2001). Evolution of generative design systems for modular physical robots. In *International Conference on Robotics and Automation*, pp. 4146–4151.

Hornby, G. S. and J. B. Pollack (2001a). The advantages of generative grammatical encodings for physical design. In *Congress on Evolutionary Computation*.

Hornby, G. S. and J. B. Pollack (2001b). Body-brain co-evolution using l-systems as a generative encoding. In *Genetic and Evolutionary Computation Conference*, pp. 868–875.

Huelsbergen, L., E. Rietman, and R. Slous (1999, September). Evolving oscillators in silico. *IEEE Transactions on Evolutionary Computation 3*(3), 197–204.

J., K. P. and T. R. Osborn (2000, 8 July). Operon expression and regulation with spiders. In A. S. Wu (Ed.), *Proceedings of the 2000 Genetic and Evolutionary Computation Conference Workshop Program*, Las Vegas, Nevada, pp. 161–166.

Jacob, C. (1994). Genetic L-system programming. In Y. Davidor, H.-P. Schwefel, and R. Männer (Eds.), *Parallel Problem Solving from Nature III*, Jerusalem, pp. 334–343. Springer-Verlag.

Jacob, C. (1996, September). Evolution programs evolved. In H.-M. Voigt, W. Ebeling, I. Rechenberg, and H.-P. Schwefel (Eds.), *Parallel Problem Solving from Nature IV, Proceedings of the International Conference on Evolutionary Computation, Lecture Notes in Computer Science*, Volume 1141, Berlin, Germany, pp. 42–51. Springer Verlag.

Jakobi, N. (1995). Harnessing morphogenesis. Technical Report CSRP 423, School of Cognitive and Computer Science, University of Sussex, Sussex.

Johnson, S. D., G. B. Parker, I. Cyliax, and D. Braun (1997). Using cyclic genetic algorithms to reconfigure hardware controllers for robots. Technical Report 494, Computer Science Department, Indiana University, Bloomington, Indiana.

Jong, K. D. (1992). Are genetic algorithms function optimizers? In R. Maener and B. Manderic (Eds.), *Proceedings of the Second International Conference on Parallel Problem Solving from Nature (PPSN)*, Amsterdam, pp. 3–13. North-Holland, Elsevier Science Publishers.

Kajitani, I., M. Murakawa, D. Nishikawa, H. Yokoi, N. Kajihara, M. Iwata, D. Keymeulen, H. Sakanashi, and T. Higuchi (1999). An evolvable hardware chip for prosthetic hand controller. In *Proceedings of the 7th International Conference on Microelectronics for Neural, Fuzzy and Bio-Inspired Systems (MicroNeuro '99)*.

Kalganova, T. (2000a, 13-15 July). Bidirectional incremental evolution in extrinsic evolvable hardware. In J. Lohn et al. (Eds.), *The Second NASA/DoD Workshop on Evolvable Hardware*, Palo Alto, CA, USA, pp. 65–74. IEEE Computer Society.

Kalganova, T. (2000b). An extrinsic function-level evolvable hardware approach. In *Proceedings of the Third European Conference on Genetic Programming (EUROGP2000), Lecture Notes in Computer Science*, Volume 1802, Edinburg, UK, pp. 60–75. Springer-Verlag.

Kargupta, H. (2000). A striking property of genetic code-like transformations. *Complex Systems Journal 13*(1), 1–32.

Kargupta, H. and B.-H. Park (2001). Gene expression and fast construction of distributed evolutionary representation. *Journal of Evolutionary Computation 9*(1), 1–32.

Karp, G. (1998). *Cell and Molecular Biology: Concepts and Experiments* (2nd ed.). John Wiley and Sons.

Kauffman, S. (1969). Metabolic stability and epigenesis in randomly constructed genetic nets. *Journal of Theoretical Biology 22*, 437–467.

Kauffman, S. (1993). *Origins of Order.* NY: Oxford University Press.

Keymeulen, D., M. Durantez, K. Konaka, Y. Kuniyoshi, and T. Higuchi (1996, October). An evolutionary robot navigation system using a gate-level evolvable hardware. In *Proceedings of the First International Conference on Evolvable Systems: from Biology to Hardware (ICES '96)*, pp. 195–209. Springer-Verlag.

Keymeulen, D., K. Konaka, M. Iwata, Y. Kuniyoshi, and T. Higuchi (1998). Robot learning using gate-level evolvable hardware. In A. Birk and J. Demiris (Eds.), *Proceedings of the Sixth European Workshop on Learning Robots, Lecture Notes in Artificial Intelligence*. Springer-Verlag.

Keymeulen, D., R. Zebulem, A. Stoica, and M. Buehler (2001). Initial experiments of reconfigurable sensor adapted by evolution. In Y. Liu et al. (Eds.), *Proceedings of the Fourth International Conference on Evolvable*

*Systems (ICES2001), Lecture Notes in Computer Science*, Volume 2210, pp. 303–313. Springer-Verlag.

Keymeulen, D., R. S. Zebulem, Y. Jin, and A. Stoica (2000, September). Fault-tolerant evolvable hardware using field-programmable transistor arrays. *IEEE Transactions on Reliability 49*(3), 305–316.

Kimball, J. W. (2002). The operon. http://users.rcn.com/jkimball.ma.ultranet/Biology-Pages/L/LacOperon.html.

Kimura, M. (1983). *The Neutral Theory of Molecular Evolution*. Cambridge University Press.

Kitano, H. (1990). Designing neural network using genetic algorithms with graph generation system. *Complex Systems 4*, 461–476.

Kitano, H. (1996a). Evolvable hardware with development. In *IEEE International Symposium on Circuits and Systems (ICAS '96)*, Volume 4, pp. 33–36.

Kitano, H. (1996b). Morphogenesis for evolvable systems. In E. Sanchez and M. Tomasinni (Eds.), *Towards Evolvable Hardware: The Evolutionary Engineering Approach, Lecture Notes in Computer Science*, Volume 1062, pp. 99–117. Springer-Verlag.

Koerner, T., U. Rueckert, and J. Sitte (1998, April). Local cluster neural network analog VLSI design. *Neurocomputing 19*, 185–197.

Koza, J. R. (1992). *Genetic Programming: On the programming of computers by means of natural selection*. Cambridge, MA: MIT Press.

Koza, J. R., D. Andre, F. H. Bennet III, and M. A. Keane (1997, April). Design of a high-gain operational amplifier and other circuits by means of genetic programming. In *6th International Conference on Evolutionary Programming (EP97), Lecture Notes in Computer Science*, Volume 1213, Indianapolis, IN, pp. 125–135. Springer-Verlag.

Kumar, S. (2004). *Investigating Computational Models of Development for the Construction of Shape and Form*. Ph. D. thesis, Department of Computer Science, University College London.

Langeheine, J., S. Foelling, K. Meier, and J. Schemmel (2000, April). Towards a silicon primordial soup: A fast approach to hardware evolution with a VLSI transistor array. In J. Miller, A. Thompson, P. Thomson, and T. Fogarty (Eds.), *Proceedings of the Third International Conference on Evolvable Systems (ICES2000), Lecture Notes in Computer Science*, Volume 1801. Springer-Verlag.

Langeheine, J., S. Foelling, K. Meier, and J. Schemmel (2001). Initial studies of a new VLSI field programmable transistor array. In Y. Liu et al. (Eds.), *Proceedings of the Fourth International Conference on Evolvable Systems*

*(ICES2001)*, *Lecture Notes in Computer Science*, Volume 2210, pp. 62–73. Springer-Verlag.

Langeheine, J., K. Meier, J. Schemmel, and M. Trefzer (2004). Intrinsic evolution of digital-to-analog converters using a CMOS FPTA chip. In *The 2004 NASA/DoD Conference on Evolvable Hardware*, pp. 18–25. IEEE Computer Society.

Layzell, P. (1999, April). Reducing hardware evolution's dependency on FP-GAs. In *Proceedings of MicroNeuro '99, 7th International Conference for Neural, Fuzzy and Bio-Inspired Systems*, CA, pp. 171–178. IEEE Computer Society.

Layzell, P. (2001, May). *Hardware Evolution: On the Nature of Artificially Evolved Electronic Circuits*. Ph. D. thesis, School of Cognitive and Computing Science, University of Sussex, Sussex.

Lee, K.-Y., D.-W. Lee, and K.-B. Sim (2000). Evolutionary neural networks for time series prediction based on L-system and DNA coding method. In *Proceedings of the Congress on Evolutionary Computation*, Piscataway, NJ, pp. 1467–1474.

Levi, D. and S. A. Guccione (1999a, September). GeneticFPGA: A java-based tool for evolving stable circuits. In J. Schewel et al. (Eds.), *Reconfigurable Technology: FPGAs for Computing and Applications, Proceedings of SPIE*, Bellingham, WA, pp. 114–121.

Levi, D. and S. A. Guccione (1999b, July). GeneticFPGA: Evolving stable circuits on mainstream FPGA devices. In A. Stoica et al. (Eds.), *Proceedings of the First NASA/DoD Workshop on Evolvable Hardware*, Los Alamitos, CA, pp. 12–17. IEEE Computer Society.

Lindenmeyer, A. (1968). Mathematical models for cellular interaction in development, Parts I and II. *Journal of Theoretical Biology 18*, 280–315.

Liu, H., J. Miller, and A. M. Tyrrell (2005a). A biological development model for the design of robust multiplier. In F. Rothlauf et al. (Eds.), *EvoWorkshops 2005, Lecture Notes in Computer Science*, Volume 3449, pp. 195–204. Springer-Verlag.

Liu, H., J. Miller, and A. M. Tyrrell (2005b, July 1–3). Intrinsic evolvable hardware implementation of a robust biological development model for digital systems. In *Proceedings of the 2005 NASA/DoD Conference on Evolvable Hardware*, Washington D.C., USA, pp. 87–92. IEEE Computer Society.

Lohn, J., A. Stoica, D. Keymeulen, and S. Colombano (2001, June). The second NASA/DoD workshop on evolvable hardware. *IEEE Transactions on Evolutionary Computation 5*(3), 298–302.

Lohn, J. D. and S. P. Colombano (1999, September). A circuit representation technique for automated cicuit design. *IEEE Transactions on Evolutionary Computation 3*(3), 205–219.

Lohn, J. D., G. L. Haith, S. P. Comombano, and D. Stassinopoulos (2000). Towards evolving electronic circuits for autonomous space applications. In *Proceedings of the IEEE Aerospace Confence*, Volume 5, pp. 473–486.

Luke, S., S. Hamahashi, and H. Kitano (1999). "genetic" programming. http://citeseer.nj.nec.com/luke99quotgeneticquot.html.

Macias, N. J. (1999a). The PIG paradigm: The design and use of a massively parallel fine grained self-reconfigurable infinitely scalable architecture. In *Proceedings of the First NASA/DoD Workshop on Evolvable Hardware*, pp. 175–180.

Macias, N. J. (1999b). Ring around the PIG: A parallel GA with only local interactions coupled with a self-reconfigurable hardware platform to implement an O(1) evolutionary cycle for evolvable hardware. In *Proceedings of the Congress on Evolutionary Computation*, pp. 1067–1078.

Mattick, J. S. (2001). Non-coding RNAs: the architects of eukaryotic complexity. *EMBO reports 2*(1), 986–991.

Mattiussi, C. and D. Floreano (2004). Evolution of analog networks using local string alignment on highly reorganizable genomes. In *The 2004 NASA/DoD Conference on Evolvable Hardware*, pp. 30–37. IEEE Computer Society.

Miller, J. F. (1999, 13-17 July). Digital filter design at gate-level using evolutionary algorithms. In W. Banzhaf, J. Daida, A. E. Eiben, M. H. Garzon, V. Honavar, M. Jakiela, and R. E. Smith (Eds.), *Proceedings of the Genetic and Evolutionary Computation Conference (GECCO'99)*, Orlando, Florida, USA, pp. 1127–1134. Morgan Kaufmann, San Francisco, CA.

Miller, J. F. and M. Hartmann (2001). Untidy evolution: Evolving messy gates for fault tolerance. In Y. Liu et al. (Eds.), *Proceedings of the Fourth International Conference on Evolvable Systems (ICES2001), Lecture Notes in Computer Science*, Volume 2210, pp. 100–111. Springer-Verlag.

Miller, J. F., D. Job, and V. K. Vassilev (2000). Principles in the evolutionary design of digital circuits – part I. *Genetic Programming and Evolvable Machines 1*(1/2), 7–35.

Milne, G. (1999, July). A model for dynamic adaptation in reconfigurable hardware systems. In A. Stoica et al. (Eds.), *Proceedings of the First NASA/DoD Workshop on Evolvable Hardware*, Los Alamitos, CA. IEEE Computer Society.

Moreno, J., J. Cabestany, J. Madrenas, E. Cantó, J. Faura, and J. Insenser (1999). Approaching evolvable hardware to reality: The role of dynamic reconfiguration and virtual meso-structures. In *Proceedings of the 7th International Conference on Microelectronics for Neural, Fuzzy and Bio-Inspired Systems (MicroNeuro '99)*.

Moreno, J., J. Madrenas, J. Cabestany, E. Canto, R. Kielbik, J. Faura, and J. Insenser (1999, July). Realization of self-repairing and evolvable hardware structures by means of implicit self-configuration. In A. Stoica et al. (Eds.), *Proceedings of the First NASA/DoD Workshop on Evolvable Hardware*, Los Alamitos, CA. IEEE Computer Society.

Ochoa, G. (1998). On genetic algorithms and lindenmeyer systems. In A. Eiben, T. Baeck, M. Schoenauer, and H. P. Schwefel (Eds.), *Parallel Problem Solving from Nature V (PPSN V)*, pp. 335–344. Springer-Verlag.

Ohno, S. (1970). *Evolution by Gene Duplication*. Berlin: Springer Verlag.

O'Neill, M. and C. Ryan (2000). Incorporating gene expression models into evolutionary algorithms. In A. Wu (Ed.), *Proceedings of GECCO 2000 Workshop on Gene Expression*, San Francisco, CA, pp. 167–173. Morgan Kaufman.

Ortega, C. and A. Tyrell (1997, November). Biologically inspired reconfigurable hardware for dependable applications. In *IEE Colloquim on Hardware Systems for Dependable Applications*.

Ortega, C. and A. Tyrell (1998, March). Evolvable hardware for fault-tolerant applications. In *IEE Colloquim on Evolvable Hardware Systems*, London.

Ortega, C. and A. Tyrell (2000). A hardware implementation of an embryonic architecture using Virtex FPGFAs. In *ICES 2000*, pp. 155–164.

Ortega-Sánchez, C. and A. Tyrell (1997, September). Fault-tolerant systems: The way biology does it! In *Proccedings of Euromicro 97*, Budapest, pp. 146–151. IEEE Computer Society Press.

Ortega-Sánchez, C. and A. Tyrell (1998, May). Design of a basic cell to construct embryonic arrays. *IEE Transactions on Computers and Digital Techniques 145*(3), 242–248.

Ozsvald, I. (1998, September). Short-circuiting the design process: Evolutionary algorithms for circuit design using reconfigurable analogue hardware. Master's thesis, School of Cognitive and Computing Sciences, University of Sussex, Sussex.

Porter, R. (2001). *Evolution on FPGAs for Feature Extraction*. Ph. D. thesis, Queensland University of Technology.

Prodan, L., G. Tempesti, D. Mange, and A. Stauffer (2001). Embryonics: Artifical cells driven by artificial dna. In Y. Liu et al. (Eds.), *Proceedings of the Fourth International Conference on Evolvable Systems (ICES2001), Lecture Notes in Computer Science*, Volume 2210, pp. 100–111. Springer-Verlag.

Prusinkiewicz, P. and J. Hanan (1989). *Lindenmayer Systems, Fractals, and Plants*, Volume 79 of *Lecture Notes in Biomathematics*. Springer-Verlag.

Prusinkiewicz, P. and A. Lindenmeyer (1990). *The algorithmic beauty of plants*. Springer-Verlag.

Ramsden, E. (2001, July). The ispPAC family of reconfigurable analog circuits. In *Proceedings of the Third NASA/DoD Workshop on Evolvable Hardware*, Long Beach, California, USA, pp. 176–181.

Ramsden, E., G. W. Greenwood, and D. Hunter (2005, July 1–3). Earp-1 – an evolvable analog research platform. In *Proceedings of the 2005 NASA/DoD Conference on Evolvable Hardware*, Washington D.C., USA, pp. 20–25. IEEE Computer Society.

Reil, T. (1999). Dynamics of gene expression in an artificial genome - implications for biological and artificial ontogeny. In D. Floreano, F. Mondada, and J. Nicoud (Eds.), *Proceedings of the 5th European Conference on Artificial Life*, pp. 457–466. Springer Verlag.

Reil, T. (2000). Models of gene regulation - a review. In C. Maley and E. Boudreau (Eds.), *Artificial Life 7 Workshop Proceedings*, pp. 107–113. MIT Press.

Roggen, D., D. Floreano, and C. Mattiussi (2003, March 17–20). A morphogenetic evolutionary system: Phylogenesis of the poetic circuit. In A. M. Tyrrell, P. C. Haddow, and J. Torresen (Eds.), *Proceedings of the 5th International Conference on Evolvable Systems: From Biology to Hardware ICES 2003*, Volume 2606 of *Lecture Notes in Computer Science*, Trondheim, Norway, pp. 153–164. Springer.

Schoenauer, M. and Z. Michalewicz (1997). Evolutionary computation. *Control and Cybernetics 26*(3), 307–338.

Seaman, G. (2000, April 2). FPGA bitstreams and open design. http://www.opencollector.org/news/Bitsream.

Sekanina, L. and R. Růžička (2003). Easily testable image operators: The class of circuits where evolution beats engineers. In *The 2003 NASA/DoD Conference on Evolvable Hardware*, pp. 135–144. IEEE Computer Society.

Shanthi, A., K. Singaram, and R. Parthasarathi (2005, July 1–3). Evolution of asynchronous sequential circuits. In *Proceedings of the 2005*

*NASA/DoD Conference on Evolvable Hardware*, Washington D.C., USA, pp. 93–96. IEEE Computer Society.

Shipman, R., M. Shakleton, and I. Harvey (2000, October). The use of neutral genotype-phenotype mappings for improved evolutionary search. *BT Technology Journal 18*(4), 103–111.

Sims, K. (1994, July). Evolving virtual creatures. In *Proceedings of SIG-GRAPH '94, Computer Graphics Annual Conference Series*, pp. 15–22.

Stauffer, A. and M. Sipper (1998). L-hardware: Modeling and implementing cellular development using l-systems. In D. Mange and M. Tomassini (Eds.), *Bio-Inspired Computing Machines: Towards novel computational architectures*, pp. 269–287. Presses Polytechniques Et Universitaires Romandes.

Stoica, A., D. Keymeulen, T. Arslan, V. Duong, R. Zebulum, I. Ferguson, and X. Guo (2004, December 6–8). Self-recovery experiments in extreme environments using a field programmable transistor array. In O. Diessel and J. Williams (Eds.), *Proceedings of the 2004 IEEE International Conference on Field-Programmable Technology (FPT'04)*, Brisbane, Australia, pp. 9–15.

Stoica, A., D. Keymeulen, R. Tawel, C. Salazar-Lazaro, and W.-t. Li (1999, July). Evolutionary experiments wih a fine-grained reconfigurable architecture for analog and digital CMOS circuits. In A. Stoica, D. Keymeulen, and J. Lohn (Eds.), *Proceedings of the First NASA/DoD Workshop on Evolvable Hardware*, Pasadena, CA, USA, pp. 76–84. IEEE Computer Society.

Stoica, A., D. Keymeulen, and R. Zebulum (2001, July). Evolvable hardware solutions for extreme temperature electronics. In *Proceedings of the Third NASA/DoD Workshop on Evolvable Hardware*, Long Beach, California, USA, pp. 93–97.

Stoica, A., R. Zebulum, M. Ferguson, D. Keymeulen, and V. Duong (2002, July 15-18). Evolving circuits in seconds: Experiments with a stand-alone board level evolvable system. In *Proceedings of the 202 NASA/DoD Conference on Evolvable Hardware*, Alexandria Virginia, USA, pp. 67–74.

Stoica, A., R. Zebulum, and D. Keymeulen (2001, July). Progress and challenges in building evolvable devices. In *Proceedings of the Third NASA/DoD Workshop on Evolvable Hardware*, Long Beach, California, USA, pp. 33–35.

Stomeo, E., T. Kalganova, C. Lambert, N. Lipnitsakya, and Y. Yatskevich (2005, July 1–3). On evolution of relatively large combinational logic circuits. In *Proceedings of the 2005 NASA/DoD Conference on Evolvable Hardware*, Washington D.C., USA, pp. 59–66. IEEE Computer Society.

Sun Microsystems Inc. (2003). Package java.lang.reflect javadoc, java 2 platform std. ed. v1.4.2. http://java.sun.com/j2se/1.4.2/docs/api/java/lang/reflect/package-summary.html.

Tempesti, G., D. Roggen, E. Sanchez, and Y. Thoma (2002). A POEtic architecture for bio-inspired hardware. In Standish, Abbass, and Bedau (Eds.), *Artificial Life VIII*, pp. 111–115. MIT Press.

Thompson, A. (1995). Evolving electronic robot controllers that exploit hardware resources. In *Proceedings of the Third European Conference on Artificial Life (ECAL95)*, pp. 640–656. Springer Verlag.

Thompson, A. (1996a). Evolutionary techniques for fault tolerance. In *Proceedings of the UKACC International Conference on Control (CONTROL '96)*, pp. 693–698. IEE.

Thompson, A. (1996b). An evolved circuit, intrinsic in silicon, entwined with physics. In *Proceedings of the First International Conference on Evolvable Systems (ICES96), Lecture Notes in Computer Science*, Volume 1259, pp. 390–405. Springer.

Thompson, A., I. Harvey, and P. Husbands (1996). Unconstrained evolution and hard consequences. In E. Sanchez and M. Tomasinni (Eds.), *Towards Evolvable Hardware: The Evolutionary Engineering Approach, Lecture Notes in Computer Science*, Volume 1062, pp. 136–165. Springer-Verlag.

Thompson, A. and P. Layzell (2000). Evolution of robustness in an electronics design. In *Proceedings of the Third International Conference on Evolvable Systems: from biology to hardware (ICES2000)*, pp. 218–228. Springer-Verlag.

Thomson, K. (1988). *Morphogenesis and Evolution*. NY: Oxford University Press.

Torresen, J. (1998). A divide-and-conquer approach to evolvable hardware. In M. Sipper et al. (Eds.), *2nd International Conference on Evolvable Systems: from biology to hardware (ICES 98), Lecture Notes in Computer Science*, Volume 1478, pp. 57–65. Springer-Verlag.

Torresen, J. (2000a). Possibilities and limitations of applying evolvable hardware to real-world applications. In *FPL*, pp. 230–239.

Torresen, J. (2000b). Scalable evolvable hardware applied to road image recognition. In J. Lohn et al. (Eds.), *Proceedings of the Second NASA/DoD Workshop on Evolvable Hardware*, pp. 245–252. IEEE Computer Society.

Torresen, J. (2001). Two-step incremental evolution of a prosthetic hand controller based on digital logic gates. In *ICES2001*, pp. 1–13.

Tufte, G. and P. C. Haddow (2000). Evolving an adaptive digital filter. In J. Lohn et al. (Eds.), *Proceedings of the Second NASA/DoD Workshop on Evolvable Hardware*, pp. 143–150. IEEE Computer Society.

Tufte, G. and P. C. Haddow (2003, 17th – 20th March). Building knowledge into developmental rules for circuit design. In A. Tyrrell, P. Haddow, and J. Torresen (Eds.), *Proceedings of the 5th International Conference on Evolvable Systems: From Biology to Hardware (ICES2003). Lecture Notes in Computer Science*, Volume 2606, Trondheim, Norway, pp. 69–80. Springer-Verlag.

Twyman, R. (2001). *Developmental Biology.* Instant Notes. Oxford: BIOS Scientific Publishers limited.

Twyman, R. (2002). *Molecular Biology* (2nd ed.). Instant Notes. Oxford: BIOS Scientific Publishers limited.

Tyrell, A. M., E. Sanchez, D. Floreano, G. Tempesti, D. Mange, J. M. Moreno, J. Rosenberg, and A. E. P. Villa (2003, March 17-20). POEtic tissue: An integrated architecture for bio-inspired hardware. In A. M. Tyrrell, P. C. Haddow, and J. Torresen (Eds.), *Evolvable Systems: From Biology to Hardware, 5th International Conference (ICES 2003)*, Volume 2606 of *Lecture Notes in Computer Science*, Trondheim, Norway, pp. 129–140. Springer.

Tyrrell, A., G. Hollingworth, and S. Smith (2001, July). Evolutionary strategies and intrinsic fault tolerance. In *Proceedings of the Third NASA/DoD Workshop on Evolvable Hardware*, Long Beach, California, USA, pp. 98–106.

Ukkonen, E. (1995). On-line construction of suffix trees. *Algorithmica 14*, 249–260.

van Remortel, P., B. Manderick, and T. Lenaerts (2004). Gene interaction and modularisation in a model for gene-regulated development. In *The 2004 NASA/DoD Conference on Evolvable Hardware*, pp. 253–260. IEEE Computer Society.

Vassilev, V. K. and J. F. Miller (2000). Scalability problems of digital circuit evolution: Evolvability and efficient designs. In J. Lohn et al. (Eds.), *Proceedings of the Second NASA/DoD Workshop on Evolvable Hardware.* IEEE Computer Society.

Vigander, S. (2001, February). Evolutionary fault repair of electronics. Master's thesis, Department of Computer and Information Science, Norwegian University of Science and Technology, Trondheim.

Vinger, K. A. and J. Torresen (2003). Implementing evolution of FIR-filters efficiently in an FPGA. In *The 2003 NASA/DoD Conference on Evolvable Hardware*, pp. 26–32. IEEE Computer Society.

Watson, R. A. and J. B. Pollack (2000). Symbiotic combination as an alternative to sexual recombination in genetic algorithms. In *Proceedings of PPSN VI*, pp. 425–434.

Watson, R. A. and J. B. Pollack (2001). Symbiotic composition and evolvability. In *Proceedings of the 2001 European Conference on ALife*, pp. 480–490.

Wee, J. W., T. S. Park, and C. H. Lee (1999). Adaptive hardware evolution under unpredictable environmental changes. In *Proceedings of the First IEEE Asia Pacific Conference on ASICs (AP-ASIC '99)*, pp. 372–375.

Winter, P., G. Hickey, and H. Fletcher (2002). *Genetics* (2nd ed.). Instant Notes. Oxford: BIOS Scientific Publishers limited.

Wu, A. S. and I. Garibay (2002, June). The proportional genetic algorithm: Gene expression in a genetic algorithm. *Genetic Programming and Evolvable Hardware 3*(2), 157–192.

Wu, A. S. and R. K. Lindsay (1996). A survey of intron research in genetics. In H.-M. Voight, W. Ebeling, I. Rechenberg, and H.-P. Schewfel (Eds.), *Proceedings of the Fourth International Conference on Parallel Problem Solving from Nature*, Berlin, Germany, pp. 101–110. Springer-Verlag.

Xilinx Inc. (1997, April 24). XC6200 Field Programmable Gate Arrays (Version 1.10).

Xilinx Inc. (2001, April). Virtex 2.5 V Field Programmable Gate Arrays: Product Specification, DS003 (V2.5). http://direct.xilinx.com/bvdocs/publications/ds003.pdf.

Xilinx Inc. (2004, September 10). Virtex-4 family overview DS112 (V1.1): Advance product specification. http://www.xilinx.com.

Yao, X. (1999, September). Evolving artificial neural networks. *Proceedings of the IEEE 87*(9), 1423–1447.

Yao, X. and T. Higuchi (1999). Promises and challenges of evolvable hardware. *IEEE Transactions on Systems, Man, and Cybernetics - Part C: Applications and Reviews 29*(1), 87–97.

Yu, T. and P. Bentley (1998, September 27-30). Methods to evolve legal phenotypes. In *Proceedings of the 5th International Conference on Parallel Problem Solving from Nature (PPSN V)*, Amsterdam. Springer.

Zebulum, R. S., D. Keymeulen, V. Duong, X. G. M. Ferguson, and A. Stoica (2003). Experimental results in evolutionary fault-recovery for field programmable analog devices. In *The 2003 NASA/DoD Conference on Evolvable Hardware*, pp. 182–186. IEEE Computer Society.

Zebulum, R. S., M. A. Pacheco, M. Vellasco, and H. T. Sinohara (2000). Evolvable hardware: On the automatic synthesis of analog control systems. In *Proceedings of the IEEE Aerospace Conference*, Volume 5, pp. 451–463.

Zebulum, R. S., A. Stoica, and D. Keymeulen (2000). The design process of an evolutionary oriented reconfigurable architecture. In *Proceedings of the Congress on Evolutinary Computation*, Volume 1, pp. 529–536.